Climate Change Education Across Disciplines K-12

Climate Change Education Across Disciplines K-12

New Jersey and Beyond

Edited By
Lauren Madden

ROWMAN & LITTLEFIELD
Lanham • Boulder • New York • London

Published by Rowman & Littlefield
An imprint of The Rowman & Littlefield Publishing Group, Inc.
4501 Forbes Boulevard, Suite 200, Lanham, Maryland 20706
www.rowman.com

86-90 Paul Street, London EC2A 4NE

Copyright © 2025 by Lauren Madden

All rights reserved. No part of this book may be reproduced in any form or by any electronic or mechanical means, including information storage and retrieval systems, without written permission from the publisher, except by a reviewer who may quote passages in a review.

British Library Cataloguing in Publication Information available

Library of Congress Cataloging-in-Publication Data

Names: Madden, Lauren, editor.
Title: Climate change education across disciplines K-12 : New Jersey and beyond / edited by Lauren Madden.
Description: Lanham, Maryland : Rowman & Littlefield, 2025. | Includes bibliographical references and index.
Identifiers: LCCN 2024040156 (print) | LCCN 2024040157 (ebook) | ISBN 9781538193310 (cloth) | ISBN 9781538193327 (paperback) | ISBN 9781538193334 (ebook)
Subjects: LCSH: Environmental education—New Jersey. | Climatic changes—Study and teaching (Elementary)—New Jersey. | Climatic changes—Study and teaching (Secondary)—New Jersey | Interdisciplinary approach in education—New Jersey.
Classification: LCC GE85.N59 C55 2025 (print) | LCC GE85.N59 (ebook) | DDC 372.35/7—dc23/eng/20241118
LC record available at https://lccn.loc.gov/2024040156
LC ebook record available at https://lccn.loc.gov/2024040157

Contents

Foreword: Climate Change Education Across Disciplines K–12 A Case Study from New Jersey ix
Tammy Murphy, First Lady of New Jersey

Acknowledgments xi

1 A Brief Overview on Climate Change Education in New Jersey and Beyond 1
  Lauren Madden

2 Effects of Climate Change in New Jersey 7
  Ian Gray

3 Teachers' Initial Reactions to Climate Change Education in New Jersey 19
  Lauren Madden, Eileen Heddy, Margaret Wang, Julia T. Sims, Samantha Lindsay, and Isabelle Pardew

4 Preparing Children for Climate Change in the Early Years 27
  Janna Hockenjos and Janice Parker

5 Climate Change Education in the Elementary Classroom 41
  Jeanne Muzi

6 Education for Climate Action in Secondary Schools 53
  Andrea Drewes and Jessica Monaghan

CONTENTS

7   Climate Change Education in the STEM Classroom   75
    Rachel DiVanno, Kelly Stone, and Melissa Zrada

8   Climate Change Connections: The Power of Knowing and
    Doing, TOGETHER!   83
    Cari Gallagher and April Oliver

9   Using Data Visualization to Enhance Climate Change Education   99
    Melissa Zrada, Kristin Hunter-Thomson, and Carrie Ferraro

10 Socioscientific Issues as a Framework for Teaching Climate Change 115
    Sami Kahn and Timothy Lintner

11 The Role of Art in Climate Change Education   131
    Carolyn McGrath

12 Climate Change and the School Library   153
    Ewa Dziedzic-Elliott

13 Sustainability and Climate Change for All: A Systematic
    Approach to Implementing K–12 Problem-Based Learning and
    Design Thinking   165
    Eddie Cohen and Brielle Kociolek

14 Incorporating Climate Emotions into the Classroom: Strategies
    to Support Student Emotional Growth and Thriving   179
    Kathleen L. Grant and Sarah Springer

15 Developing an EcoJustice Consciousness for Classroom Teachers
    and their Students   193
    Marissa E. Bellino and Greer Burroughs

16 High-Quality Instruction Begins with Great Resources   211
    Beverly Plein, Julia T. Sims, and Margaret Wang

17 Bridging Classrooms and Communities for Climate Change
    Education   223
    Tina Overman, Kelly Stone, Chris Turnbull, and Karen Woodruff

18 Nonformal Education about Climate Change   237
    Pat Heaney, Allison Mulch, and Graceanne Taylor

19 "Mom, What's for Dinner?" Climate Action Served with a Side of Hope 249
    Helen and Grace Corveleyn

20 Effective Climate Change Education in the Classroom Begins
    with Effective Climate Change Professional Learning for Educators 261
    Karen Woodruff, Missy Holzer, and Andrea Drewes

21	Educational Policy & Climate Change Education in New Jersey *Sarah Sterling-Laldee*	279
22	Case Studies from Outside New Jersey *Christine Whitcraft, Kelley Lê, Kimi Waite, Jim Clifford, and Amara Ifeji*	285
23	Climate Change in the Garden State's Science Standards *Glenn Branch, National Center for Science Education*	299

| Index | 303 |
| Contributor Biographical Sketches | 319 |

Foreword

CLIMATE CHANGE EDUCATION ACROSS DISCIPLINES K–12 A CASE STUDY FROM NEW JERSEY

Tammy Murphy, First Lady of New Jersey

Every August, thousands of stingrays congregate along the shoreline of Island Beach State Park, gliding along the azure water as waves crest and rush up the soft, white sand. From abundant farm stands to deep, wild forests, preserved wetlands and mountain summits, I could name countless idyllic images, all local to the Garden State of New Jersey. Because of the threat of a warming climate, however, these places and habitats, as well as our way of life, are at risk.

In response to this risk, we harnessed one of our state's greatest resources—our public education system—and in June 2020, we incorporated climate change across seven of our nine K–12 student learning standards, with the remaining two standards finalized in October 2023. This addition means that New Jersey is the first state in the nation to ensure our students are prepared across multiple content areas to both preserve all that we love about the Garden State and to succeed in an economy that will undoubtedly be changed by the climate crisis.

In each grade from kindergarten through senior year of high school, our students are looking at climate change from every angle and preparing to become the next generation leaders of climate literate policy makers, data analysts, entrepreneurs, urban planners, researchers, anthropologists, economists, artists, and more.

For this, we are grateful to our outstanding educators, who keep New Jersey consistently ranked among the very best for our state's public education system. Importantly, that gratitude must also come with the essential professional development educators need to incorporate these new standards into their classrooms. It is our hope that this text will support that endeavor. Many of the authors in this book are educators themselves and leaders in New Jersey's climate change education movement and have included personal anecdotes and sample lesson plans to inform and guide those who are new to this space.

Our approach to climate change education marks a seismic shift in the way protecting the environment has long been viewed. Rather than an impediment, we understand that New Jersey's transition to a green economy will serve as a once-in-a-generation opportunity for growth, which we are undertaking with an intention to achieve equity and environmental justice. Our students' education is a powerful tool in this work, as it will allow our students to think critically about their future role in the green economy during their most formative years.

Ultimately, incorporating climate change education across our learning standards has the potential to build wealth across all communities, increase job opportunities in brand new industries, empower our entrepreneurs, improve our residents' health, and preserve the beauty of our state, all through the power of education.

With so much to gain and no time to lose, we hope this educational movement will progress quickly across the nation, and thanks to New Jersey's educators, our students will be leading the way.

Acknowledgments

This book was the result of many hardworking and brilliant individuals who said, "yes," when asked to share their best ideas and strategies for teaching about climate change. It has been an incredible honor to collaborate with each member of the author team to stitch together this collection. When New Jersey became the first state to include standards supporting climate change instruction across grade levels and subject areas, we knew this was a historic moment. By telling the stories of "what works" and "how to" as we begin to support teachers and children learning about our changing planet, we hold the light a little higher for others following in our path.

In particular, I'd like to express my deepest gratitude for New Jersey's First Lady Tammy Murphy. Mrs. Murphy's bold advocacy for climate change education elevates the voices of children and ensures their perspectives are central as we build solutions to this global challenge. Many thanks to the dedicated educators across New Jersey who have worked tirelessly in schools, parks, museums, nature centers, and beyond to prepare these children with big ideas and hopeful mindsets. To the activists, scientists, artists, creators, and storytellers inspiring us all throughout this climate journey: thank you.

Nathan Davidson, Hollis Peterson, and their wonderful team at Rowman & Littlefield/Bloomsbury have been exceptional supporters throughout the entire process of bringing this book to fruition. Kerry Rushnak provided critical logistical and organizational support in the early stages of writing this book. Tara Ronda's expert advice, edits, and suggestions improved the book immensely. Korie Vee's cover art brings these stories to life.

To my students, colleagues, and administrators at TCNJ: Thank you for showing me that the best way out of any predicament is through creativity and innovation. To my family, especially Connor and Luke, I'm the luckiest to have your support and love. —*Lauren Madden, September 2024.*

Individual chapter authors wish to extend further thanks to the following:

- Many thanks to Alexandra Vargas and Emma Casler for going above and beyond as first-year teachers, finding innovative ways to implement EcoJustice education in their classrooms. We would also like to thank all of the other EcoJustice educators for sharing their experiences. —*Marissa Bellino & Greer Burroughs*
- I would like to extend a special acknowledgment to the children of New Jersey who continually inspire us with their curiosity and innovation in improving the world around them. A heartfelt shout-out to Shane, Troy, and Hope—your creativity and passion are paving the way for a brighter future! —*Eddie Cohen*
- The educators at Newtown Public Schools, TCNJ, and Metuchen Public Schools, and especially her family for their passion for learning. —*Rachel DiVanno*
- With thanks to Valerie Frost-Lewis and all the children and teachers at Peppermint Tree Child Development Center. —*Janna Hockenjos*
- Appreciation to the Nature Based Education Consortium Climate Education Advocacy Working Group. —*Amara Ifeji*
- With appreciation for the entire team of the ECCLPS project. —*Kelley Le & Christine Whitcraft*
- Thanks to the student artists in my classes at Hopewell Valley Central High School who have so generously shared their art as well as their thoughts and feelings about climate change and the youth climate activists, such as Nyombi Morris, who have, through example, shown the path forward for young people around the world. —*Carolyn McGrath*
- Many thanks to Teacher Leaders, Kristin Hopson and Stacey Moore who took a glimmer of an idea about climate learning and crafted an amazing program, and to all the Slackwood School Climate Kids, past, present, and future. —*Jeanne Muzi*
- With thanks to all of the staff and students at Bear Tavern Elementary School for continuing to inspire us! —*Tina Overman and Chris Turnbull*
- Thanks to the following teachers for contributing vignettes: Angelique Hammack, Tina Ezzo, Kim Talarico, and Mike Skomba. —*Julia T. Simms, Beverly Plein, and Margaret Wang*
- Appreciation to Roger Zuidema and Mary Leou. —*Sarah Sterling-Laldee*
- With gratitude to the students and staff at George L. Cantebrone Elementary School. —*Kelly Stone*

CHAPTER 1

A Brief Overview on Climate Change Education in New Jersey and Beyond

Lauren Madden

In June 2020, New Jersey announced that it would become the first state in the United States to adopt comprehensive standards to support climate change learning across subject areas and grade levels.[1] This announcement, under the leadership of New Jersey First Lady Tammy Murphy, was made during the early part of the COVID-19 pandemic, when schools were facing unprecedented uncertainty and teaching and learning looked vastly different from years past. Even so, the announcement also sparked a nationwide conversation about the importance of preparing young children for the challenges that they will inevitably face in the future.

The launch of these standards was announced soon after New Jersey's Department of Environmental Protection released its *2020 Scientific Report on Climate Change*,[2] which delineated the ways New Jersey is affected by climate change. In many cases, these changes, such as rising temperatures and sea levels, affect New Jersey at a disproportionately higher rate than other parts of the globe. Not surprisingly, policies and plans put in place to respond to these changes and prioritize innovations that help us mitigate the effects of climate change followed suit. For example, Governor Murphy launched the statewide Office of Climate Action and the Green Economy in 2020 and championed several initiatives focused on bolstering green energy through wind and solar power.[3] It follows that ensuring New Jersey can develop and maintain a workforce equipped to support these green initiatives is a key priority. Thus, climate change education also became a priority.

Though climate change education can build the future workforce for a green economy, it should also be viewed as a potential solution, sparking individual and collective actions in learners. Some preliminary research also supports the value of preparing children for learning about climate change. For example, using mathematical models, Cordero and colleagues[4] found that "the implementation of climate change education over a 30-year period (2020–2050) could reduce emissions by 18.8 GT of CO_2, an amount that would rank in the top quarter (15 out of 80) of the presented solutions in Project Drawdown." Although this assertion is speculative, it suggests tangible differences in actions and behaviors can be sparked in learners who are aware of the magnitude of the crisis our planet faces.

It should also be noted that the June 2020 announcement was not the first step toward comprehensive climate change education in New Jersey. Through its regular evaluation and adoption of standards across subject areas, New Jersey adopted the Next Generation Science Standards[5] (NGSS) as its science learning standards in 2014. These standards, referred to as performance expectations, use a three-dimensional approach to learning in that science content is always presented within the context of science and engineering practices, or the ways scientists and engineers do their work, and crosscutting concepts, or the overarching ideas that cross scientific disciplines. The three-dimensional performance expectations are all situated within the bigger "storyline" of learning. Each performance expectation is presented in a table where the disciplinary core ideas (science content), science and engineering practices, and crosscutting concepts are specified. The standards that come before, during, and after a grade level are also highlighted, allowing for vertical articulation of ideas across schools and school systems. Finally, the tables include explicit connections to the Common Core State Standards in English-language arts and mathematics, allowing teachers to purposefully plan interdisciplinary investigations. The NGSS also added several new science and engineering concepts to the suite of standards, namely global climate change. However, this content is included only at the middle school and high school levels of the NGSS.

So, at the time of the June 2020 announcement, New Jersey schools had already begun addressing climate change in science classes at the middle and high school levels and had created the initial roadmap for interdisciplinary learning through the NGSS framework. This first step eased the way for adoption into other content areas and grade levels for certain, but further guidance was needed to ensure our state's introduction of these important changes was successful. To aid this process, the Sustainability Institute at The College of New Jersey and the New Jersey School Boards Association convened a group of thought leaders to generate suggestions for best practices in climate change education implementation. This convening included school leaders, higher education faculty, educators from nonprofits, and other stakeholders from across the state. The thought leaders met regularly electronically and were asked to share suggestions and guidance using an anonymous online survey. They were also asked to share the survey link with several knowledgeable colleagues who could provide additional feedback and suggestions. They presented a summary of the survey results in light of existing research regarding climate change education in the *Report* on *K-12 Climate Change Education Needs in New Jersey*.[6] The report included a number of suggestions that fell into several large categories, including providing professional learning opportunities for teachers, providing access to high-quality curricular materials, ensuring support from administrators and school boards, and creating connections with community organizations.

The standards to support learning about climate change for all subjects outside of English and mathematics were officially approved by the New Jersey Board of Education and released to the public in the spring of 2022; teachers began implementing them in the fall of 2022. Several months into the 2022–23 academic year, the state announced that grants would be available for schools to support their teachers and students in implementing these new standards, and many schools opted to apply for this supplementary funding, which was used in a multitude of ways, including for professional learning for teachers and curricular materials. The English and mathematics standards were introduced in the spring of 2023 and were eventually voted upon and approved by the state school board in the fall of 2023. Classrooms across the state will begin implementing these standards in the 2024–25 academic year.

Where Does New Jersey Fit in the Bigger Picture?

It should be clear that climate change education has been prevalent in New Jersey, across the country, and around the world for quite some time. In her seminal book *Saving Us*,[7] Katharine Hayhoe describes the work of scientists who caution society about the risks of atmospheric warming due to greenhouse gas emissions at the turn of the twentieth century. Similarly, professional organizations such as the North American Association for Environmental Education and the National Science Teaching Association have published position statements on the importance of teaching about climate change for several decades. Nonprofits such as the Climate Reality Project and the World Wildlife Foundation have also supported large-scale climate education. Likewise, many teachers, schools, and districts have elected to implement climate actions within their classrooms, buildings, and communities. Bryce Coon, a seasoned high school teacher, shared with *Education Week* why he left the classroom to focus his professional efforts entirely on the climate crisis, expanding the scope of his work beyond his own classroom and school.[8]

Required teaching about climate change can shift the impact and influence of individual teachers significantly. Yet, when it comes to formalized efforts at scale, New Jersey is the first in the United States to make climate change education a requirement. In fact, globally, New Jersey also sits near the earliest adopters of climate change education for all students across subjects and grade levels. Italy remains the only country where comprehensive climate change education is compulsory for all students, having implemented this policy about a year ahead of New Jersey.[9] Other states like Maine, California, and Connecticut are also ahead of the curve with statewide implementation of climate change education, which will be detailed more clearly in Chapter 22 of this book.

About This Book

The goal of this edited volume is to share recommendations from teachers, higher education faculty, informal educators, and scientists doing great work to ensure schools have the tools to effectively implement climate change instruction across grade levels and disciplines. The remainder of the front matter covers the specific effects of climate change on New Jersey and findings from some preliminary research on teachers' readiness to adopt these new standards. The bulk of the book provides strategies, success stories, and tried-and-true suggestions for teachers and school leaders. In conclusion, we offer recommendations for continued professional learning, policy implications, and shared stories from other states whose journeys into climate education look different from New Jersey. Together, the chapters in this edited volume represent a wide-spanning, though not exhaustive, scope on climate change education that can hopefully be used to inform others.

Notes

1. "First Lady Tammy Murphy Announces New Jersey Will Be First State in the Nation to Incorporate Climate Change Across Education Guidelines for K-12 Schools," New Jersey Office of Governor, June 3, 2020. https://nj.gov/governor/news/news/562020/approved/20200603b.shtml.

2. New Jersey Department of Environmental Protection, *New Jersey Scientific Report on Climate Change* (Trenton, NJ: State of New Jersey, 2020), https://dep.nj.gov/wp-content/uploads/climatechange/nj-scientific-report-2020.pdf.

3. "About Governor's Office of Climate Action," New Jersey Office of Climate Action & the Green Economy, 2024. https://nj.gov/governor/climateaction/about/

4. Eugene C. Cordero et al., "The Role of Climate Change Education on Individual Lifetime Carbon Emissions," *PLOS One* 15, no. 2 (2020): 1–23, https://doi.org/10.1371/journal.pone.0206266.

5. See https://nextgenscience.org.

6. Lauren Madden, *Report on K-12 Climate Change Education Needs in New Jersey* (Trenton, NJ: New Jersey School Boards Association & Sustainable Jersey, February 2022), https://www.njsba.org/wp-content/uploads/2022/02/climate-change-ed-online-2-2.pdf.

7. Katharine Hayhoe, *Saving Us: A Climate Scientist's Case for Hope and Healing in a Divided World* (New York: Simon & Schuster, 2021).

8. Bryce Coon, "I Quit Teaching to Become a Climate Activist. Here's Why." *Education Week,* November 29, 2023. https://www.edweek.org/teaching-learning/opinion-i-quit-teaching-to-become-a-climate-activist-heres-why/2023/11.

9. Gianluca Mezzofiore, "Italy to Become First Country to Make Learning About Climate Change Compulsory for School Students," CNN, November 6, 2019. https://www.cnn.com/2019/11/06/europe/italy-climate-change-school-intl-scli-scn/index.html.

CHAPTER 2

Effects of Climate Change in New Jersey

Ian Gray

New Jersey is a state with diverse landscapes, all of which are riddled with signs of things that have changed and things that have remained the same over millions of years. Twenty-five thousand years ago, the last ice sheet covering the northern portion of the state left incredible variability and richness behind as it retreated, once again, to the Arctic.[1] The coastline has shifted and changed with ocean levels, and constantly shifting tides leave behind small changes daily. The Highlands and Piedmont regions have been shaped by shifting tectonic plates over millions of years, creating a landscape of swales, mountains, ponds, and stream corridors, and the Pine Barrens have grown up and burned down again and again in a cycle that would seem endless on a human time frame. All these small and dramatic changes recurring over time have sculpted the state into what it is today. Human impacts on this cycle of change, particularly the acceleration of it, cannot be overstated. Climate change has impacted New Jersey in countless ways. To best understand these impacts, it makes sense to break them out into categories, though it is important to remember that all the affected systems are interconnected. Impacts of climate change can be seen and felt in the air, in water, and on land. If we consider these three categories and the way they affect humans and the natural ecosystems around us, we can start to scratch the surface of what climate change is changing in New Jersey.

Air: Change Rooted in the Sky

The warming air temperatures across the globe are the most-discussed aspect of climate change. Greenhouse gases, the compounds in the atmosphere that trap heat and create the warming effect that allows life on Earth, have increased in concentration dramatically since the Industrial Revolution and the advent of fossil fuel exploitation. In New Jersey, the trend of warming average temperatures holds true, as it does globally, with five-year average temperatures increasing nearly two degrees Fahrenheit since 1895.[2] New Jersey is warming at a faster average pace than the rest of the northeast[3] and is already feeling the effects of warmer temperatures. Warming average temperatures pose risks to wildlife, native plants, and the people of New Jersey. A particularly at-risk stakeholder is the state's agriculture industry. Extreme summer heat can be deleterious to livestock and decrease overall crop production during the peak growing season.[4] Additionally, fruit orchards are at risk of losing entire crops to inconsistent spring temperatures, which can cause early flowering followed by cold snaps that kill flower buds. Warming winter temperatures also put the blueberry and cranberry growers of the Pinelands at risk due to the plants' cold winter temperature requirements. Both fruits are staple crops for the economy and cultural cornerstones of the Pinelands. New Jersey is among the top ten producers of both in the United States. Thus, rising air temperatures will lead to more impacts on New Jersey agricultural systems than only frost damage and harvest declines.

Warmer air can hold more moisture than colder air. Year-round warmer temperatures will lead to higher humidity levels and dryer soils as hot air pulls moisture from the ground and contributes to changing precipitation patterns. During the spring and summer months, consistently warmer temperatures contribute to longer and more frequent periods of drought in some areas, with more severe rain in others. Just like hotter growing season temperatures, periods of drought affect crop productivity and increase energy use in agriculture; less consistent rain means more energy must be used for crop irrigation. Outside the agricultural industry, higher humidity levels also influence everyone living in New Jersey. Of course, hot and humid summer days are not a new phenomenon in New Jersey; however, longer and stronger heat waves have become more common. Heat waves are particularly dangerous for New Jersey residents since so much of our state's landmass has been developed and paved. Impermeable surfaces like asphalt and concrete hold heat longer than natural surfaces and don't allow cooling overnight, resulting in a feedback loop of hotter and hotter days in developed areas and a higher likelihood of heat-related injury for people living in urbanized areas. Heat-related injury is one of a multitude of human health hazards New Jersey faces with climate change.[5]

The famous "hole in the ozone layer" that forms over Antarctica when the natural level of ozone high in the stratosphere drops well below the accepted threshold due to atmospheric pollution brought ozone to the forefront of environmentalism in the mid-1980s.[6] Stratospheric ozone levels have been of concern all over the globe, not just over Antarctica, because the ozone layer importantly filters ultraviolet (UV) radiation from the sun. This issue has seen some successful remediation over the last fifteen to twenty years as more stringent regulations have been placed on polluters. Now, we face a different kind of ozone crisis: increases in ground-level ozone, which is an air pollutant and can be dangerous to humans. Ozone is a major focus of the New Jersey Department of Environmental Protection (NJDEP) because it is a pollutant that can continue to increase in ground-level concentrations with warming temperatures even without any additional human input.[7] So, even as more stringent restrictions to air pollution are put in place as air temperatures continue to rise, ground-level ozone will still be a concern.

One of the most concerning processes involved in increasing ozone levels in New Jersey and elsewhere is the feedback loop caused by higher emissions generated from fossil fuel–driven power plants placed under higher load during heat wave events. When heat waves hit, the demand for electricity to cool indoor spaces rises and the efficiency of power plants decreases, which generates more greenhouse gas emissions and leads to higher ground-level ozone concentrations.[8] In New Jersey, the state's focus on atmospheric air quality over the last twenty years has made some promising impacts on ground-level ozone. The eight-hour-daily average of ozone pollution measured in New Jersey dropped from almost 0.08 parts per million (ppm) in 1997 to 0.07 ppm in 2022, and the average of the highest daily annual ozone concentrations documented across the state has also consistently dropped over the last twenty years.[9]

Ozone and humidity levels are two ways climate change impacts air quality. Since the summer of 2023, the type of air pollution at the forefront of New Jerseyans' minds is without a doubt wildfire smoke. Throughout the summer, raging wildfires in Canada burned over forty million acres, with major impacts to air quality throughout the entire northeastern United States.[10] Wildfire smoke is classified as an air pollutant due to the fine particulate matter (particles smaller than 2.5 microns that are particularly damaging to the human respiratory system) it produces. Though wildfires happen in the Pine Barrens and elsewhere within the state and cause small spikes in air pollutants, they cannot compare to the smoke that travels into the state from larger wildfires on the west coast and to the north.[11] More frequent and intense wildfires now occur due to the suppression of fire in ecosystems resulting from European settlement in North America, which has allowed large quantities of fuel to build up in forests that historically had less intense fires on a more regular basis to keep fuel loads in check. As summer months become increasingly dryer and hotter, wildfires will continue to grow in frequency due to easier ignition conditions. The New Jersey Forest Fire Service and many local government agencies and nonprofits perform more frequent prescribed burns, fires intentionally set and controlled under carefully selected conditions, throughout the state to reduce fuel loads and provide ecosystem services to native species that rely on fire. Controlled burns, seen in figure 2.1, do produce smoke and particles 2.5 micrometers or smaller that pose risks to the human respiratory system; however, the controlled nature of the practice allows more control over smoke direction and preliminary public notification, which is not available for wildfires.

Water: Too Much and Too Little

Annual rainfall amounts are increasing slightly, with extreme events increasing more dramatically and periods of drought becoming more common, as seen in figure 2.2. Aquifers, the underground reservoirs of clean water from which we draw for household and commercial use across the state, will be stressed as draw volume continues to increase and extended periods of drought continue. Heavy rainfall events interspersed with periods of little to no rainfall cannot recharge aquifers the same way a historically "normal" distribution of rainfall can.[12] New Jersey gets over seventy percent of its drinking water from surface water, which is also at risk of increased contamination levels with more common flood events. While other human-introduced contaminants like agricultural waste runoff and wastewater from manufacturing pose a major risk to clean drinking water, the rising temperature of surface water will increase risks of other climate change–induced water contamination.

Harmful algal bloom (HAB) events continue to rise in frequency in New Jersey surface waters. These events are stimulated by warm water temperatures and nutrient overload in water from agricultural runoff that contains high levels of phosphorus and nitrogen. HABs are not actually algae at all; they are caused

Figure 2.1 A prescribed burn at Mercer Meadows in Mercer County, NJ. Controlled fires like these are set by the New Jersey Forest Fire Service in partnership with landowners and managers. (Ian Gray, Mercer Meadows RxB, 2021)

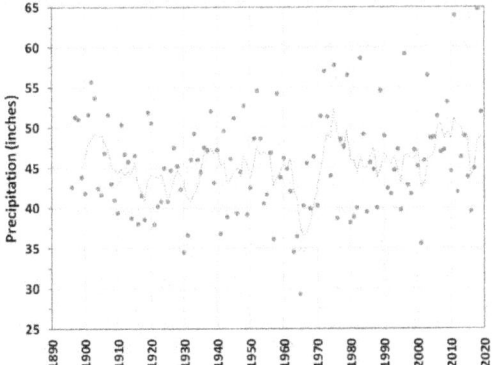

Figure 2.2 Statewide Annual Precipitation in Inches (1895–2019). Points represent the statewide annual precipitation, and the dashed line represents a five-year average of the data based on year of interest and the previous four years. Data acquired from the Office of the New Jersey State Climatologist, 2020 (New Jersey Department of Environmental Protection, New Jersey Scientific Report on Climate Change, Version 1.0 [Trenton, NJ: NJDEP, 2020], 38)

by different types of cyanobacteria that can produce dangerous cyanotoxins. These toxins are harmful to cattle, pets, and people and can cause fever, vomiting, and in extreme cases of high exposure levels, death. The NJDEP uses a community science-driven plan to solicit the reporting of suspected HABs, with confirmation by DEP scientists. In 2022, the number of suspected HAB events reported increased by over sixty percent, which caused the number of lab-confirmed HAB events in the state to increase more than eighty percent.[13] HAB events are increasingly likely to reoccur in a body of water once they have happened, and they will continue to increase in frequency as temperatures rise.[14] No HAB-related human deaths have ever been recorded; however, there have been multiple cases of HAB-related illness and as frequency increases at both recreational water bodies and in sources of drinking water, the risk of more human illness increases.

With severe storms increasingly common, erosion (the process of soil and silt sediments being removed by water from an area) and sedimentation (the process of soil and silt sediments moved by erosion being deposited elsewhere) are increasing across the state's aquatic ecosystems. This is particularly problematic as stream-bed erosion tends to function in a positive feedback loop, where during high-water events, the stream cuts further and further into the ground, causing channelization and separating the stream from the floodplain, not allowing the energy of moving water to dissipate and causing further erosion in future flood events. As this cycle continues, the stream bed cuts deeper during high-water events and gets lower into the ground, which not only lowers the level of surface water but effectively lowers the water table, as well. Because of this erosion and water-table lowering, it becomes harder for plants to access groundwater to grow and more difficult for wildlife to find easily accessible water.

As streambank erosion and other types of erosion occur, there is a secondary problem: sediment deposit downstream. All the dirt, sand, silt, and debris moved by fast-paced water is eventually deposited elsewhere and can clog culverts under roads and train tracks; fill in reservoirs used for drinking water; and degrade habitat for fish, amphibians, and aquatic insects that rely on clean stream bottoms to live. Additionally, in coastal areas, storm surge, the dramatic increase in sea level caused by strong, low-pressure weather systems, combined with flooding from rainfall leads to incredibly destructive flood events. Superstorm Sandy in November 2012 is likely the most famous example of this in New Jersey, where the storm surge and rainfall combined with a full moon and high-tide event, creating one of the most destructive flooding events in New Jersey history, causing $29.4 billion in damages.[15] Up to $8.1 million of the damage caused by Hurricane Sandy can be attributed to the increased vulnerability of coastal communities caused by human-influenced sea-level rise.[16] Between the landfall of Hurricane Sandy in 2012 and September 2017, the State Hazard Mitigation Plan identified another seven major coastal erosion incidents: hurricanes, tropical storms, nor'easters, and other storms that caused significant flood damage and coastal erosion in the state. This is only a small percentage of major storms the state experienced in that timeframe; however, it demonstrates the frequency of storms causing major damage along the coast.

In addition to erosion in freshwater systems and along the coast, rising sea level and warming ocean temperatures lead to another major issue for saltwater ecosystems in the state: ocean acidification. As atmospheric carbon dioxide concentrations rise, ocean waters rapidly dissolve CO_2 and the pH of the water decreases, increasing the acidity of ocean water. Approximately one-third of human-caused CO_2 emissions are dissolved into the ocean, so the connection between climate change and ocean acidification is clear. More acidic water impacts oceangoing creatures that rely on calcification of exoskeleton structures, the most common example being shellfish. A recent study also found that diatoms, plankton with silica-based exoskeletons that were once thought to be advantaged by acidification may not benefit as once thought.[17] Because silica breaks down slower in more acidic water, the nutrient recycling of silica from decomposing diatoms on the ocean floor has slowed significantly. This may lead to a lack of available silica for new diatoms to form exoskeleton structures at the highly productive surface levels of the ocean, where diatoms are the most abundant organism harnessing energy from the sun by photosynthesis, and therefore are one of the cornerstones of the ocean's food web. New Jersey has a prevalent fishing and aquaculture industry, contributing over $1 billion to the state economy annually, but these sectors are at risk due to the lowering pH of oceans. Though ocean pH is most dramatically affected by increasing atmospheric CO_2, as levels continue to rise, freshwater ecosystems will also be impacted by acidification.[18]

Sea-level rise is another threat to the state, independent of the high water levels caused by storms and damage from flood events. Rising sea levels will impact communities and economies along the Jersey shore, as well as in metropolitan areas. A less direct impact will occur in the communities along the Delaware River system as far north as Trenton, where tidal marshes and river levels are strongly influenced by ocean water levels and tides.[19] The impacts of rising sea levels on tidal ecosystems will increase due to the amount of human impact already in these sensitive areas: Impoundments, dikes, and drainage ditches have been installed in a majority of tidal wetlands throughout the state for agricultural purposes and wildlife conservation. Though these structures are needed for many reasons, they will dramatically limit wetlands' ability to compensate for rising water levels in both fresh and saltwater marsh ecosystems, which will change ecosystem function and may lead to a decrease in plant and wildlife diversity.[20]

Land: Small but Mighty

The systems of our environment are complex and interdependent, and some of the greatest changes to the land of New Jersey happen because of changes to our hydrologic systems. As mentioned, erosion is a critical issue in the state as climate change continues to increase the frequency of severe rainfall events. New Jersey's moniker of "The Garden State" speaks to the thriving farming and nursery industries in the state. Erosion is a massive issue for growing crops, as nutrient-depleted soils more easily wash away during heavy rainfall events. Additionally, changing precipitation patterns mean more frequent droughts that stress farm and garden crops alike. Many New Jersey farmers have taken on these challenges by adopting sustainable agricultural techniques like no-till farming to reduce soil erosion and water conservation strategies like supporting natural wetlands to conserve water and decrease the need for irrigation.

It could be argued that the greatest loss New Jersey is experiencing and will experience due to climate change is the loss of native biodiversity. An emerging threat to our state as the average temperature climbs is the emergence of new invasive species of plants and animals.[21] Native biodiversity is an important asset in any region, and New Jersey is particularly interesting in this sense due to the state's varied ecoregions. The stark contrast from the Ridge and Valley region in the northwestern part of the state to the Atlantic Coastal Plain in the southeast and everywhere in between can be shocking to those unfamiliar with how diverse New Jersey really is. This ecological diversity is a major asset, and we are at risk of losing most of the remaining natural diversity in these areas to climate change. Kudzu vine is a primary example of the danger an invasive species can pose as the climate warms. Famously known as "the vine that ate the south," kudzu is native to southeast Asia and was introduced to the United States at scale during the 1930s by the Civil Conservation Corps, a government work program started by President Roosevelt to fight erosion in the Dust Bowl–affected south while stimulating the Depression Era economy by employing young American men.[22] Once introduced, the vine quickly established in the southern states and aggressively took over wood edges, forests, and even buildings. Though kudzu can survive in colder temperatures, it grows aggressively in places with mild winters and hot, wet summers. As of 2023, fewer than forty populations of kudzu have been identified in New Jersey, and the New Jersey Invasive Species Strike Team has taken action to eradicate as many as possible before it spreads.[23] This effort will help slow or even stop the vine's spread in our state; however, as the climate warms and our winters become milder, we are likely to see kudzu and other harmful invasive plants take over. Invasive plants are an issue for agriculture, industry, and residential areas. The displacement of native species in preserved wild areas is one of the biggest concerns about invasives in New Jersey, though. As invasive plants expand their range, they cut down the available area in which native plants and animals can live in a state where so much land is already either heavily developed or farmed.

Aquatic invasive species are just as, if not more concerning than terrestrial invasives. New Jersey's aquatic ecosystems are under threat due to drought, warming temperatures, and acidification. Many sensitive species struggle to survive and cannot handle the increased pressure of aquatic invasive species taking resources from them. Aquatic plants such as water chestnut, watermilfoil, and hydrilla grow aggressively and can easily choke out shallow-to medium-depth water, creating dense patches of vegetation that make small bodies of water uninhabitable for fishes and choke out native submerged aquatic vegetation that plays a key role in ecosystem functions. Invasive fishes like the flathead catfish and northern snakehead have the potential to outcompete native species as well as economically important sport fish. Aquatic invasives have the potential to spread quickly throughout watersheds and can move rapidly on recreational equipment like kayaks, motorboats, and fishing gear.

Invasive species taking over native habitat is not the only concern for New Jersey's biodiversity due to climate change. The primary issue facing our native species is the changing average and extreme temperatures within current native ranges. New Jersey falls within an ecological transition zone, with many species associated with southern US habitats in the southern part of the state and many species associated with more northern US habitats in the northern region. In other words, our state is at the southernmost and northernmost range for many species.[24] As the climate changes and we see an overall trend of warming temperatures, New Jersey is likely to lose many "northern" species that are at the southern terminus of their range in the state.

At the other end of this shifting habitat equation is a glimmer of hope. This process of shifting habitats regarding climate is happening everywhere, meaning that species displaced from New Jersey may find refuge in states north of us, and simultaneously, New Jersey itself will be able to provide refuge to some species losing their habitat to the south, thus undergoing a shift in species distribution. The risk of losing biodiversity remains, though, as many species are unable to shift their range at the same pace as the temperature changes. This is particularly true of trees, which are slow to spread from population sources due to their slow growth. Additionally, most plants in New Jersey produce seed in summer or fall, which means that any plants needing birds to distribute their seeds through consumption have a difficult time expanding their range northward due to the fall migration of birds moving predominantly in a southern direction. These challenges lead to conservation organizations actively moving species northward through a process called assisted migration.[25] There are, of course, risks to this type of approach, but the potential risks of losing species native to New Jersey and the US overall due to climate change force ecologists' hand.

Conclusion

Climate change is impacting the entire world. Some of the impacts are the same regardless of location, while others are more particular to a region. New Jersey is being impacted by just about every negative aspect of climate change: warming temperatures, droughts, floods, severe storms, and ocean acidification, to name a few. Though New Jersey has been at the confluence of multiple habitat types and geological formations for millions of years, the current rate of change in the ecological, hydrological, and meteorological systems of the state is unprecedented. Human-caused climate change is the reason for this. Luckily, New Jersey's government agencies, including the DEP, are among the leaders in the United States in addressing climate change with policies and legislation, grant funding, and education. New Jerseyans are an incredibly proud group of people, from Newark and Newton to Cape May and Cinnaminson. The people here, like our ecosystems and industries, are resilient and willing to fight for the things that matter to us all. With this book, the impact of New Jersey educators empowering future generations of New Jersey citizens as staunch defenders of sustainability and conservation will only increase. The voices of those students fighting against climate change will be the greatest impact it has had on New Jersey yet.

Notes

1. Scott D. Stanford et al., "Chronology of Laurentide Glaciation in New Jersey and the New York City Area, United States," *Quaternary Research* 99 (2021), 142–67. doi:10.1017/qua.2020.71.
2. New Jersey Department of Environmental Protection, *New Jersey Scientific Report on Climate Change*, Version 1.0 (Trenton, NJ: NJDEP, 2020), 32.
3. Richard Horton et al., "Northeast," in *Climate Change Impacts in the United States: The Third National Climate Assessment*, ed. J. M. Melillo, Terese Richmond, and G. W. Yohe (Washington, DC: U.S. Global Change Research Program, 2014), 16-1-nn.
4. "Farming, Food, and Climate Change in New Jersey," New Jersey Climate Change Resource Center, May 2020, https://njclimateresourcecenter.rutgers.edu/climate_change_101/farming-food-and-climate-change-in-new-jersey/.
5. "Farming, Food, and Climate Change"
6. Steve Colwell and Jonathan Shanklin, "The Ozone Hole," British Antarctic Survey, June 30, 2022, https://www.bas.ac.uk/data/our-data/publication/the-ozone-layer/.
7. NJDEP, *Report on Climate Change*, 23.
8. NJDEP, *Report on Climate Change*, 62.
9. New Jersey Department of Environmental Protection, *Cyanobacterial Harmful Algal Bloom (HAB) Freshwater Recreational Response 2022 Summary Report* (Trenton, NJ: NJDEP, April 2023).
10. David Wallace-Wells, "'It's Like Our Country Exploded': Canada's Year of Fire," *The New York Times*, October 24, 2023, https://www.nytimes.com/2023/10/24/magazine/canada-wildfires.html.
11. NJDEP, *Report on Climate Change*, 62.
12. NJDEP, *Scientific Report on Climate Change*, 73.
13. NJDEP, *Cyanobacterial Summary Report*.
14. NJDEP, *Report on Climate Change*, 78.
15. New Jersey Office of Emergency Management, "Coastal Erosion and Sea Level Rise," in *New Jersey State Hazard Mitigation Plan 2019* (Trenton, NJ: NJOEM, January 25, 2019), 5.2.
16. Benjamin H. Strauss et al., "Economic Damages from Hurricane Sandy Attributable to Sea Level Rise Caused by Anthropogenic Climate Change," *Nature Communications* 12, no. 1 (2021): 2720, doi:10.1038/s41467-021-22838-1.
17. Jan Taucher et al. "Enhanced Silica Export in a Future Ocean Triggers Global Diatom Decline, "*Nature* 605 (2022): 696–700. doi:10.1038/s41586-022-04687-0.
18. "What Is Ocean Acidification? And How Will It Affect New Jersey?" New Jersey Climate Change Resource Center, September 2020, https://njclimateresourcecenter.rutgers.edu/climate_change_101/ocean-acidification/.
19. NJDEP, *Report on Climate Change*, 79.
20. "Sea Level Rise in New Jersey: Projections and Impacts," New Jersey Climate Change Resource Center, May 2020, https://njclimateresourcecenter.rutgers.edu/climate_change_101/sea-level-rise-in-new-jersey-projections-and-impacts.
21. New Jersey Department of Environmental Protection, *New Jersey's State Wildlife Action Plan* (Trenton, NJ: NJDEP, 2018).

22. Nancy J. Loewenstein et al., "History and Use of Kudzu in the Southeastern United States," Alabama A&M and Auburn University Extension, March 8, 2022, https://www.aces.edu/blog/topics/forestry-wildlife/the-history-and-use-of-kudzu-in-the-southeastern-united-states/.

23. Michael Van Clef et al., "Vine That Ate the South Comes North," New Jersey Conservation Foundation, August 21, 2019, https://www.njconservation.org/vine-that-ate-the-south-comes-north.

24. Tom Gilbert et al., "The Intersection of Wildlife and Climate Change," New Jersey Conservation Foundation, February 17, 2022, https://www.njconservation.org/the-intersection-of-wildlife-and-climate-change/.

25. Stephen Handler et al., "Assisted Migration" USDA Forest Service Climate Change Resource Center, 2018, https://www.fs.usda.gov/ccrc/topics/assisted-migration.

CHAPTER 3

Teachers' Initial Reactions to Climate Change Education in New Jersey

Lauren Madden, Eileen Heddy, Margaret Wang, Julia T. Sims, Samantha Lindsay, and Isabelle Pardew

In September 2022, teachers across New Jersey began to implement standards to support climate change learning. To provide a snapshot of teachers' preparedness to integrate climate change into their instructional practices both before and after implementing these standards, The College of New Jersey (TCNJ), in partnership with SubjectToClimate, sent a survey to teachers across the state in June and December 2022.[1] It is important to note that an earlier version of this study appears on the SubjectToClimate website.[2] The key takeaways of this study were that, at the times of the survey administration, there was insufficient integration of climate change into the existing curriculum, lack of professional development on the topic, uncertainty regarding the interdisciplinary nature of climate change education, and a general reporting of the efficacy of the New Jersey Climate Change Education Hub (the Hub).

The Survey

Teachers were recruited to participate in the survey via emails on professional listservs. First, teachers employed at schools within TCNJ's Professional Development School Network (PDSN) were contacted. The PDSN is a network of twenty-five school districts—all within a thirty-mile radius of TCNJ—that offer field experiences for teacher candidates and provide feedback on various programs. After being sent to PDSN district administrators, who were encouraged to share it with their staff, the survey was distributed more broadly throughout the state using an email list maintained by TCNJ's School of Education. This list contains contact information for past participants in professional development sessions offered by the School of Education. The same procedure was followed for the June and December survey administrations.

The survey first asked demographic questions regarding school district, teaching assignment, subject, and grade level(s) taught. Next, respondents were asked to assess their familiarity with the standards to support climate change learning, confidence in teaching about climate change, current teaching practices, climate change education professional development experiences, and resources used to teach about climate change. Respondents were then asked a series of questions about their school and community readiness to address climate change, modeled on a similar study administered to teachers throughout the European Union in May–June 2020.[3] At the end, teachers were provided space to note further questions and comments.

Survey Respondents

In the June 2022 administration, fifty-one individuals responded to the survey from at least twenty-three different school districts.[4] In the December 2022 administration, seventy-one individuals responded, representing at least thirty-five different school districts. Just under eighty percent of respondents indicated that they worked as teachers, with the remaining percentage split between other school professionals, including counselors and administrators.

Secondary educators at the middle or high school levels represented the largest category of respondents, about three-quarters of the pool, with relatively fewer teachers responding from the elementary level. With respect to subject, it might not be surprising that science comprised the highest percentage of respondents, at around half, though other subject areas such as social studies, mathematics, and English-language arts were also represented. The subjects with lower percentages of representation included career readiness, physical education, visual arts, and world languages.

Preparedness to Integrate Climate Change Education into the Curriculum

It should be noted that in this chapter we'll discuss overall trends without detailed data analyses. A detailed reporting of the data can be found in the white paper on SubjectToClimate's website.[5]

The respondents were asked to indicate using a five-point Likert-type scale their familiarity with the standards to support climate change learning, familiarity with the New Jersey Department of Education's (NJDOE's) resources on climate change education, and confidence integrating climate change into their instruction. Though there was a small increase in teachers' feelings of preparedness between the two survey administrations, the difference was not statistically significant. The smaller June 2023 dataset reflected a continuation of the trends observed in the first two surveys: Educators have become more familiar with the updated learning standards and are gradually gaining confidence in their ability to integrate climate change into the classroom.

Current Integration of Climate Change into Instruction

Respondents were asked to report how often they integrated climate change into their instruction in the June and December surveys. The largest group of respondents, approximately a third, indicated they integrate climate change occasionally in class discussions. A smaller number of responses indicated more frequent climate change instruction through either a full instructional unit or a series of lessons, with only a handful of respondents suggesting they integrated the topic throughout the entire school year. About 15 percent of respondents maintained they did not integrate climate change at all. Findings were quite similar among the smaller June 2023 dataset, but notably, no respondents in this administration indicated they did not teach about climate change, hopefully a benefit of the updated New Jersey student learning standards. However, we should be careful about drawing too many generalized conclusions from this dataset given its relatively small sample size.

Professional Learning

The surveys asked teachers how much professional development they had received on climate change education. They were also asked about the format of any professional development that had occurred. In both surveys, just over half of the respondents reported having little to no professional development on the subject, while fewer than one in ten described taking at least one course or unit at the undergraduate or graduate level. The remainder had mixed experiences, ranging from a few hours to a module or unit in an undergraduate or graduate course. Similar trends were revealed in the June 2023 survey, although we observed a notable increase in those who had pursued full-day professional development opportunities.

Respondents were asked to describe their professional learning experiences in terms of its delivery (e.g., in-person or online, self-paced or self-taught). The bulk of respondents reported their experiences were asynchronous and self-paced. However, most teachers indicated they preferred in-person professional development opportunities. With each survey administration, an increasing number of respondents sought out courses at higher education institutions, suggesting they were interested in developing a more comprehensive understanding of the content related to climate change.

Instructional Resources

The survey asked teachers how they had found or adapted curricular resources for teaching about climate change in the 2022–23 academic year. They were allowed to select multiple responses including those on the NJDOE website, existing instructional materials from previous years, professional organizations like National Science Teachers Association or National Council of Teachers of Mathematics, media sources such as National Geographic or PBS, or the Hub created by New Jersey Climate Change Education Initiative. Overall, the highest number of respondents, about two-thirds, reported using the NJDOE's resources and about half cited media sources. Respondents from the June 2023 survey reported higher usage of the Hub—nearly doubling the initial reported percentage—along with more interest in resources from professional organizations related to their subjects. Other trends remained approximately the same as in the first two surveys.

Similarities to Prior Studies

School Education Gateway, an organization funded by the European Commission, conducted a climate education survey from May to June 2020 and received responses from thirty-six countries. Our study used modified items from the EU study to better contextualize NJ teachers' experiences relative to others globally. Both our study and the EU study concur that most teachers believe regional curricula do not sufficiently address climate change education.[6] Teachers were asked if they felt schools in their regions were equipping students with knowledge and skills to understand climate change and take appropriate action in their own lives. Over 80 percent of respondents in each survey maintained that school curricula do not sufficiently address climate change education.

Respondents were asked to select reasons teachers might not include climate change in their lessons. While just over half of the EU respondents believed climate change was outside their subject area, a greater percentage reported there was a lack of expertise/training among teachers in their schools or regions. Our data mirrors the EU data in that the biggest reason for excluding climate change is lack of expertise and training, along with a lack of climate change education resources related to the educator's subject area.

In our June 2022 survey, about three-quarters of respondents believed teachers did not have the necessary knowledge to teach about climate change, which concurs with the responses from prior questions. However, despite this belief, most respondents maintained that climate education *can* be integrated into subjects other than science. Overall, respondents did not believe the curriculum sufficiently addressed climate change or discussed climate action. In December 2022, after about one semester of integrating climate change education into the curriculum, participants' responses were similar to the June 2022 survey. Our findings align with the EU data, as there is agreement on the imperative to include climate change education in the curriculum but a reluctance to embrace the interdisciplinary nature of the subject. In addition, the level of disagreement is higher when it comes to assessing the current implementation of climate change knowledge and climate action into learning standards. In our small June 2023 follow-up, teachers' attitudes followed the trend observed from the June to December 2022 period: There was more agreement that climate change can be an interdisciplinary subject despite more skepticism that teachers are prepared to integrate climate change across all subjects.

Understanding Similarities and Differences

Overall, we saw very few differences in the data between the three time periods we studied. To better understand the data, we set out to hypothesize factors responsible for variation among respondents. Several survey questions were analyzed based on those respondents who had reported using the Hub created in partnership with SubjectToClimate, who adapted their platform to create a database for New Jersey educators with resources on interdisciplinary climate change education. We sought to examine differences in responses given by those who had reported using the Hub versus those who did not. While we noted some intriguing trends, it is important to note these differences are not derived from large sample sizes. The sample sizes for the June 2022 and December 2022 surveys were fifty-one and seventy-one, respectively. Such numbers cannot ensure statistical significance for findings.

When asked about their familiarity with the standards to support climate change learning, three-quarters of respondents who had used the Hub rated their knowledge between a 3 and a 5, on a scale of 1 to 5, suggesting Hub users were at least somewhat familiar with the standards. Hub users also reported having experienced more professional development on climate change and more in-depth professional learning experiences (e.g., multiple workshops) than those who had not used the Hub. There was also a difference between Hub users and non-users regarding frequency of teaching about climate change. In both 2022 surveys, more than ninety percent of Hub users had integrated climate change into their curriculum at least once, which was ten percent more than non-Hub users. Perhaps this could be due to the approximately inverse trends of the training and learning opportunities—the percentage of Hub users who had at least some professional development was nearly equal to the percentage of non-Hub users who had no professional development. These trends imply that more awareness of the Hub could prove to be very helpful to New Jersey teachers. When we consider the smaller dataset from June 2023, it appears that teachers overall are integrating more climate change instruction into their practice, and Hub users are doing so even more frequently and with increased confidence.

Conclusion

Overall, New Jersey educators believe climate change to be an essential component of interdisciplinary education, but there is a lack of sufficient resources for professional development and teacher training on the subject. We observed few differences between respondents in different survey cohorts, indicating that teachers did not feel adequately prepared to implement New Jersey's updated standards and that they did not iterate these updates over the course of this survey timeline. The preference for in-person professional development indicates a need for more learning opportunities to be made accessible to New Jersey educators. Additionally, there is a need for more access to climate change education resources that support teachers in identifying appropriate climate change integrations for their specific grade levels and subject areas. Given the promising trends we found among the small cohort of respondents that used the Hub, perhaps this resource could be a mechanism for closing this gap.

Future Work

Since the trends related to Hub users are not based on large sample sizes of educators, future studies should focus on evaluating the efficacy of the New Jersey Climate Change Education Hub, observing whether respondents answered questions any differently after consulting with it. This would help evaluate available resources and determine suitability in relation to what educators are looking for. In addition, if the study was modified to follow the same set of teachers over the course of a semester, the data may provide insight into how educators' viewpoints changed—or remained the same—after a certain period. This would allow us to further investigate whether educators believe they have adequate resources, professional development, and network support, while also assessing their views on the implementation of climate change education in New Jersey.

Notes

1. A follow-up survey was sent to teachers in June 2023, and this administration included far fewer participants. We'll comment generally on changing trends based on the June 2023 data but will not use these data as part of the larger chapter.

2. Lauren Madden et al., *New Jersey Teachers' Perspectives on the Implementation of K-12 Climate Change Standards* (SubjectToClimate, 2023), https://subjecttoclimate.nyc3.cdn.digitaloceanspaces.com/files/nj-climate-change-standards-implementation-paper.pdf

3. "Survey on Climate Education – Results," European School Education Platform, July 31, 2020, https://school-education.ec.europa.eu/en/insights/viewpoints/survey-climate-education-results.

4. During both administrations, several respondents did not provide a district.

5. Madden et al., *New Jersey Teachers*.

6. "Survey on Climate Education."

CHAPTER 4

Preparing Children for Climate Change in the Early Years

Janna Hockenjos and Janice Parker

As climate change becomes a reality and education standards that support learning about climate change make their way into our elementary and high schools, educators and decision makers are discovering that opportunities to integrate environmental education across all subjects abound. Yet, we continue to navigate our own feelings and fears to bring awareness to this important aspect of living and learning in our K-12 communities. Where do we start? Many question if this content is developmentally appropriate for our youngest learners.

How young is too young to teach children about climate change? In the past couple of years, with the uptick in natural disasters and displaced communities, the argument that children are too young to learn about climate change is losing steam. For many young people, it's their daily reality. The real question we need to ask is, how do we approach this content? Climate change is a complex issue for most adults to understand, let alone our youngest children. Yet, as educators, it is critical we ensure that children in the early years have opportunities to explore and discover what climate change is and how they can make a difference in the world. This begins by creating an environment where young children can grow to love the natural world around them and learn the effects of human impact through the built world around them. Teaching climate change to children this young means teaching them the basics of where and how we live, and like with any other early education content, play, developmentally appropriate practice (DAP), and social-emotional learning (SEL) play significant roles in climate education.

It is important that adults, too, understand the basics of the way our natural and built environments function. In fact, we must acknowledge the adult misconceptions of climate change educators may face, one example being the difference between weather and climate. Weather is the state of the atmosphere on any given day, whereas climate is a collection of weather conditions at a place over a period of time. Another common mistake is using the terms *global warming* and *climate change* interchangeably. Global warming is specific to atmospheric temperature changes, while climate change describes the long-term statistics of temperature, pressure, and winds sustained over several decades or longer and is the result of several occurrences, such as external factors, natural or internal processes, and anthropogenic (human-driven) forces. Of particular importance are the "anthropogenic forces," which are the variables we have the opportunity to control in order to lessen the climate crisis. Herein lies possibility. We can show even our youngest learners that their choices and actions can have a positive human impact. They can be a force for good.

When it comes to sorting out the basic facts of climate change, educators can rely on a multitude of resources for quality information, like national organizations (i.e., the Nature Conservancy, National Wildlife Federation) and state-level resources (i.e., Audubon Societies, conservation groups), as well as climate educational hubs (i.e., Project Learning Tree, Project WET, the National Oceanic and Atmospheric Administration, and SubjectToClimate). We are also starting to see environmental education curricula for early education in programs like Earth Friends and Tinkergarten, all of which can help educators surmount one major hurdle in teaching climate change to young children: preparedness. Many teachers want to teach, but many also do not feel prepared. One EdWeek Research Center survey found that "26 percent of teachers said they haven't talked about climate change because they can't think of any way it is related to the subject they teach. Nine percent said they think their students are too young to learn about it."[1]

However, unlike some of the focus in K-12 education, the goal of early childhood education is not so much *retention* of knowledge as much it is an *introduction* to it. At this early stage of academia, climate or environmental education can be viewed as an educational infrastructure. Having our youngest learners discern right from wrong or telling them how to think and feel about the state of our world further exacerbates the current environmental, sustainability, and equity challenge: how to instill inherent and intrinsic behaviors of awareness and preservation in human beings. In the years to come, children will be met with complex natural and humanitarian challenges, opportunities to act, and a deeper understanding of their impact and that of their peers, but at this formative age, environmental education is successful solely as introduction and observation.

Similar to how language starts with an understanding that letters exist, we can give children the understanding that planetary systems, ecological relationships, and production cycles exist. What they do with that information is far less important than knowing the information is out there. While most preschool children cannot use letters to spell words, some trace letters, some organize letters to spell their name, some sing their ABCs. Now, imagine a child entering first grade with no familiarity with letters. How might that child feel at school or interact with their peers? It is possible the child would fall behind academically, not engage as much socially, and significantly lack a means to express themselves and their talents.

New Jersey is the first state in the nation to include standards that support climate change learning across all content areas and in all grade levels. These standards were enacted in September 2022 and create the need for a comprehensive implementation plan across all schools in New Jersey. The climate-aligned standards for math and literacy were voted on by the New Jersey State Board of Education in October 2023 and will be enacted soon.

The *Report on K–12 Climate Change Education Needs in New Jersey* identified environmental literacy in early education as a key need, stating, "Explicit attention should be paid to foundational experiences in preschool learning environments."[2] The New Jersey Department of Education states,

With these new K–12 education standards and the voluntary inclusion of environmental content in elementary and high school classrooms and curricula across the country, there is an increasing need for young children to have a foundation of content-based environmental literacy education that includes and goes beyond nature-based, child-led experiences or outdoor open-ended play.

We have often heard the word "literacy" used in reference to language skills (the ability to read and write) and more recently with math (competence or knowledge in a specific area), and now we can include "environmental literacy," which supports climate change education. According to the North American Association for Environmental Education (NAAEE), environmental literacy is "more than what you know."[3] An environmentally literate person, as defined by the NAAEE and National Science Foundation is "someone who, both individually and together with others, makes informed decisions concerning the environment; is willing to act on these decisions to improve the well-being of other individuals, societies, and the global environment, and participates in civic life."[4] Climate change is a global challenge, and environmental literacy is a global opportunity to unite a generation beyond race, gender, and geography.

Developmentally Appropriate Practice for Early Childhood

Early childhood years are a child's most formative years—when habits form, behaviors develop, and neuroplasticity is in our favor. It is therefore imperative to integrate climate change education and environmental awareness and literacy into the early childhood education curricula with respect for DAP and SEL. Psychologist Jean Piaget shows us that young children are "more concrete thinkers," that children at these ages learn based on what they can see, hear, feel, and experience, so we know lessons are best experienced when they include hands-on activities and modeling behaviors, when material is covered through open-ended questions and play, and that collaboration in learning environments is key.[5]

Perhaps most important in these early years is not only that we strip the complex concepts down to their most basic, but that we approach every subject through the lens of possibility and positivity, and as co-learners. When adults discuss climate change, the conversation is rarely joyful or optimistic. Climate change brings out our fear, our anger, and our differences. Seeing our destroyed natural world is heartbreaking, climate anxiety is real, and political views feel inescapable. Standards that support climate change learning do not dismiss these realities; however, educators must consider that fear, anger, and anxiety is not yet the story of young children, and along with adhering to DAP and SEL, we must be mindful not to project our feelings about climate change into our classrooms. Climate change education must begin at the most basic level of learning and discovery, where each content-based lesson builds and advances from that point.

Educators, too, benefit from approaching climate content as learners. Adults as co-learners can instill a sense of hope and hold space for children to think critically, ask questions, and have conversations at a young age about this complex topic. Several methods to engage children on the topic of climate change have proven successful in primary research, many of which currently exist within most early childhood curricula:

- Ask questions that allow children to explore all angles of a subject through engaged discussions, such as, "What does a whale need to be safe?" or, "Why will it help make the world a better place?"
- Create hands-on activities and model behaviors, such as sorting items together that would typically be tossed in the trash sight unseen, gathering recycling, or turning compost to promote experience-based learning.
- Teach topics through movement to develop gross and fine motor skills through song, sign language, and repetitive motions.
- Demonstrate science experiments, such as water wheels and solar ovens, so children can form and share opinions based on observation.
- Make time for creative activities to encourage interpretation of concepts and expression through art.
- In pre-K and early elementary classrooms, reading books such as *Old Enough to Save the Planet*, *Little Blue Planet*, *Finding My Wild*, *Me and My Sit Spot*, and *Zonia's Rainforest* help children connect to what they might be able to do to make the world a better place.
- Use stations or centers where groups of children can learn more about a variety of different environmental issues, such as saving energy, reducing water waste, and recycling by having discussions.
- Make posters of what students are learning to hang around the school.

When teaching, all methods of interpretation must be accepted and discussed, including a response to a question that may be categorically or statistically wrong. Each answer a child gives provides a pallet for thinking about and exploring subjects and scenarios. Very rarely do educators need to say, "No, that is not right" or "That is not or true." Occasionally, educators can ask, "Why do you think that?" Educators may counter what is known to be false information with questions, but when we limit our teaching to what is right and wrong, what exists and what does not, we restrain the future creators and innovators who will play a role in humans surviving and possibly thriving through the climate crisis.

Not only must we meet young children where they are with developmentally appropriate lesson materials, but we must also embed climate change education throughout a young child's daily routines, interactions, and activities within the context of their impact on the world around them (their school, home, neighborhood) and within their realm of comprehension. The school or community's integration allows a young child to engage in these new-to-them behaviors and develop a community of like-minded young peers. These children, who have been given a sound foundation and understanding of their natural and built environments and their impact on it, will naturally evolve into children who have been permitted the self-agency to be an important part of their world. These children will create positive anthropogenic forces. A successful climate change education curriculum for early childhood education must include the following components:

- Integrate earth science, biodiversity, ecology, equity, and systems thinking into the current curricula;
- Build on existing climate studies (like Creative Curriculum's Reduce, Reuse, Recycle) and nature school philosophies (like that of the NAAEE);
- Address the early education needs for DAPs, SEL, whole-child development, play-based learning, and community collaboration;
- Prepare young children to continue the pathway to environmental and climate education designed for older children; and
- Foster community.

More so than solely climate change education, we need children to think critically about becoming climate conscious and resilient without the added stress and fear society may put on them. Educators need to foster settings where children develop environmental awareness, learn "green skills," and form sustainable habits. The less we talk about this topic, the scarier it becomes. Stories can empower young people to ask questions and have conversations. Curricula and specifically designed lesson plans can embolden teachers to show up with the materials to become co-learners with the next generation. Community events can spark collaboration between schools, children, families, and local organizations.

But where can a teacher start? Some recommended resources for teachers to use to strengthen their own background knowledge and enhance their classroom instruction as well.

Recommended Resources

BOOKS

- *All We Can Save: Truth, Courage, and Solutions for the Climate Crisis*, edited by Ayana Elizabeth Johnson and Katharine K. Wilkinson
- *Braiding Sweetgrass: Indigenous Wisdom, Scientific Knowledge and the Teaching of Plants*, by Robin Wall Kimmerer
- *Under A White Sky: The Nature of the Future*, by Elizabeth Kolbert
- *The Carbon Almanac: It's Not Too Late*, by The Carbon Almanac Network
- *The Parrot and the Igloo: Climate and the Science of Denial*, by David Lipsky
- *Miseducation: How Climate Change is Taught in America*, by Katie Worth
- *The Climate Optimist Handbook: How to Shift the Narrative on Climate Change and Find the Courage to Choose Change*, by Anne Theresa Gennari
- *Saving Us: A Climate Scientist's Case for Hope and Healing in a Divided World*, by Katharine Hayhoe
- *Parenting in a Changing Climate: Tools for Cultivating Resilience, Taking Action, and Practicing Hope in the Face of Climate Change*, by Elizabeth Bechard

Lesson Plan for Children Ages 3–6

EARTH FRIENDS CURRICULUM: MODULE 1, LESSON 4[6]

As we begin with our earliest learners, the lesson plan below is an example of how climate change education and environmental awareness was taught to three classrooms with students ranging from ages 3 to 6 to highlight specific examples of how the children engage with the material and experiences. To further integrate the twenty-minute lesson materials throughout the week, there are supplementary (and highly recommended) teacher activities and family activities to follow the classroom lesson.

Earth Friends Curriculum
Module 1: Our Planet
Lesson 4: Connect

PREPARING YOUNG CHILDREN 33

Objective: To comprehend the effects of temperature changes on the three elements on earth: land, water, and air
Vocabulary: Atmosphere
Reading: *Here We Are*
Science activity: Mr. John & the Polar Bear

BEGIN LESSON

[Move, 5 minutes]

Our Solar System

This is a repeated activity with American Sign Language (ASL) signs for each lesson of Our Planet. The repetition will help the children learn that earth is a planet in our solar system, and it is special because it supports life. We live on earth!

For the activity, one child or educator will represent each of the eight planets and our sun. If there are more children than astrological objects, use this as an opportunity to introduce taking turns: "We will do this again and different friends can be planets." Or offer that some children can be the moon or the stars we see in the sky at night. Use the images in We Love Our Planet often to familiarize the children with the planets and views from space.

We are going to learn about the different planets in our solar system. Every planet is special in its own way, and we are going to pretend to be the planets moving around the sun.

- SUN: *Our Sun gives us light during the day and all the energy we need to live.* Make a C-motion away from forehead with right hand.
- MERCURY: *Mercury is the smallest planet.* Squeeze two fingers together to indicate tiny.
- VENUS: *Venus is the brightest planet.* Use two hands to fan out from center.
- EARTH: *Earth is our home and the only planet that supports life.* Cross arms in front of chest for hug.
- MARS: *Mars has the biggest volcano.* Form a triangle with index fingers and then make an explosion sign with right hand.
- JUPITER: *Jupiter is the biggest planet.* Move hands apart from the center.
- SATURN: *Saturn is the planet with rings.* Draw a ring around right hand with left.
- URANUS: *Uranus is the coldest planet.* Shiver.
- NEPTUNE: *Neptune is the farthest away.* Point far away with right index finger.

[Sit & discover, 7 minutes]

CQ: What happens when ice gets hot? What does a blanket do?

Allow the children to discuss how blankets work and what happens to ice; remember, their answers do not have to make sense.

[Main lesson]

*Earth's **atmosphere** is special because it keeps earth's temperature just right for life. It's like the perfect blanket, and it's made out of gases. But right now, some humans are making choices that make too many gases and make the blanket too thick. Earth is getting too hot, and the atmosphere is getting too thick. We don't want the land or the water or the air to get too hot here on Earth. Let's see what happens when Earth's temperature gets too hot.*

[Move & watch, 8 minutes]

Mr. John & the Polar Bear

For this activity, set up a scenario where on one side, a human figure (Mr. John) is on land with a home and on the other side, a polar bear figure (the polar bear) is on ice. Make sure you can see water in between the ice and the land. Let the children observe.

Materials needed:

- Large shallow bowl or deep baking tray
- Large chunks of ice (freeze ice on a shallow dessert-sized plate)
- Figurines: human, dog, polar bear, Arctic animal, ocean animal
- Small house or building figurine
- Flat rocks or a similar object
- Hot water*
- Cold water

Preparation: In the bowl or tray, position the flat rocks on one side and place the human (Mr. John) and his house and dog on the rocks. On the other side of the bowl or tray, position the ice and place the polar bear on top. Fill the bowl or tray with cold water until the water is level with Mr. John's land (but not flowing over). Note: Small ice chunks will float, which is why they need to be big. This is not to scale! Place the ocean life figurine anywhere in between.

The polar bear lives in the polar climate zone, and Mr. John lives in the temperate climate zone. Let's see what happens when earth's temperature gets too hot. I have some hot water here.

Now, pour hot water into the existing water so the ice melts. As the ice melts, you can see the water level (sea level) rise, where Mr. John is on land at the same time you see the polar bear lose the ice to stand on. (You have the option to use a hairdryer to heat the air around the ice in place of pouring in hot water. Use what will work best for the children, as the sound and experience of a hairdryer is different than that of pouring water. However, either will melt the ice.)

What do you see happening? How is Mr. John? The Polar Bear?

Let the children watch ice melt under a polar bear and see the water level rise to the level of Mr. John's home.

Now two earth friends don't have a safe home.

Closing sentiment: *When the temperature of earth's atmosphere gets too hot from too many gases, it can affect everyone's homes. Ice can melt, and sea water can rise. Near an ocean, we can see the level of the ocean get higher. We don't want earth's blanket to get too thick.*

[END LESSON]

Teachers in the Classroom

TEACHER ACTIVITY

Melting Ice Cube Race

Materials needed:

- 2 plates (not paper)
- 2 ice cubes
- 1 glass jar
- A sunny spot

Place one ice cube on each plate. Place a jar over one ice cube. Set both plates in a sunny spot. Show the children the ice cubes and talk about how one is covered with a jar.

What do you think will happen? Which one will melt first?

Then do an activity for about 7–10 minutes and return to the plates with the children to see which ice cube is melting faster.

Which ice cube is melting faster? I wonder why.

Talk about how the sun heats the ice cube in the jar faster, because the jar, like Erth's atmosphere, traps the heat.

Throughout the Week

- Discuss: What happens when the temperature on Earth gets too hot
- Hands-on: Take the Classroom Earth down and help the children take turns placing the climate zone markers on it. Remind them with each one what that climate is like.
- Create: Continue to glue fabric to the Classroom Earth to finish the mosaic.
- Reading: *Here We Are*
- Vocabulary: Earth, natural element, climate zone, atmosphere

Good choices:

- I can learn about what kind of weather we have each season.
- I can ask an adult what happens when it gets too hot in my climate zone.

Families at Home

Discover: We learned that earth's atmosphere is like the perfect blanket, but there are some things humans are doing that are making the blanket too thick, and earth is getting too hot. Here at home, you can experiment with what happens when the blanket is too hot. You'll need a few blankets and some time. This is a good activity to do while reading a book or watching a movie together.

First, get the perfect blanket and cuddle up. Talk about how you're not too cold and you're not too hot. Now, get a thicker blanket or layer more than one blanket together. Really wrap up and get toasty warm. See how long it feels good with a blanket that's too thick. *What do you feel like when you can't take it off? Whew. It's hot.* (No one needs to get sweaty. This is just a chance to understand when the layers are too thick, the temperature goes up. Alternatively, you can do this activity with a sweatshirt or sweater.)

Visit: a greenhouse. You might know someone with a greenhouse, or you can go to a local home and garden store, which has a greenhouse section. Notice how the temperature feels inside and outside of the greenhouse.

Explore: Go outside on a rainy day. Take a walk in the rain. Get an umbrella and rainboots and see what nature is like when it's wet!

Vocabulary:

- Earth—**Earth** *is the only planet where we humans, animals, and plants can live. This makes Earth very special. Earth is not too hot and not too cold. It's just right.*
- Natural element—*Earth has three* **natural elements** *that make it a good home for humans like you and me: land, water, and air. Animals and plants need land, water, and air, too.*

- Climate zone—*Earth has three **climate zones**. The polar zone is cold, and the tropical zone is hot. In the temperate zone, the weather changes every season. In spring, it's cool and the rain helps the plants grow. In summer, it's sunny and hot and we can play in the water. In fall, it's cool and the trees lose their leaves. In winter, it's cold and sometimes snowy. Our zone is unique because our seasons are so different.*
- Atmosphere—*Earth's **atmosphere** is special because it keeps earth's temperature just right. It's like the perfect blanket, and it's made of what we call gases. But right now, some humans are making choices that make that blanket too thick, and earth is getting too hot. This is called global warming. We don't want the land or the water or the air to get too hot here on earth.*

Closing Thoughts

To love our home, we have to understand its function. It would be far easier to jump headfirst into BIODIVERSITY! It's colorful and engaging to learn about animals and plants, trees, and sea creatures. Even mushrooms are cooler than climate zones. But the truth that we can share with young children—even, especially, preschool children—is that life doesn't exist without certain conditions. As it turns out, when you add colors, hands-on activities, movement and motions, young children can begin to grasp the concepts of hot and cold places on earth; weather patterns beyond their own window; and the importance of land, water, and air here on our goldilocks planet.

The lessons shared here are only a snapshot of what the children at Peppermint Tree Child Development Center have experienced in the last two years with increased attention to climate change education, even at the preschool age. Integrating a foundation of environmental awareness and literacy into their education and activities has invigorated new interests, sparked their self-agency, and created unique solutions to solve climate problems. True, their solutions are probably not the answers, but the fact that four-year-olds are exploring this pathway of thinking and innovating is a thousand steps in a new direction.

Before you walk through the red doors and into the classrooms, you'll pass by two big blue barrels. These barrels are the reason for the newfound delight on rainy days at Peppermint Tree. You can hear that *tink tink tink* of the barrel filling up. Around here, collected precipitation is like gold compared to water from the hose that hangs in loops to the right of the barrels. No one knew the barrels would be such a hit, or perhaps it is truer to say no one knew if they'd be more trouble than fun. But in two years, not one little person has tried to knock one over, which is not too hard when the barrels are empty, and only a couple of curious minds have drained them to watch the water splash, because water splashing is always fun.

Instead, the barrels are respected for their natural resource—water for the garden. The barrels are a sort of garden lesson enhancement. The school already had a garden and a few activities scheduled around planting and harvesting, but the rain barrels serve as a greater effort to create a foundation of environmental literacy for the children at Peppermint Tree Childhood Development Center. Rain barrels, vermicomposting, perennial fruiting bushes, bird houses, a new outdoor classroom space, and soon a wildflower meadow. Still, the crux of their environmental education is not happening outdoors.

Open a bright red door to enter any classroom, and you will find tacked to the wall a circle, four feet in diameter, made of a mosaic of green, blue, and white fabric, a collective art project of the children. "That's our earth!" a little boy exclaims, our planet's name sounding more like something that rhymes with surf. "See it's blue, for water. I put that one on." He points to a piece of blue ribbon the size of a dime that he glued off the coast of the Yucatan.

The Classroom Earth is a collaborative piece of art that doubles as a hands-on learning tool for the children's lessons on everything from climate zones to animal habitats. The children spent a week gluing bits of fabric onto a stenciled outline of the North American side of the globe. The land is filled with shades of green ribbon pieces, the ocean with blues, and there are intermittent swathes of white for clouds with more purposeful collections of white fabric to cover the North and South Poles.

Through this art, earth—our unique goldilocks planet—has become less abstract. It has worked its way into their brains as the representation of the place we live, a place that we don't want to get too hot. "We don't want earth to get too hot, because then the Chico can't play," another child points out when attention is drawn to the Classroom Earth. "We want to keep earth's blanket just right."

It appears that every science experiment is an opportunity to name something. When the children were witnessing what happened to Mr. John and the polar bear, the experiment to show what happens when the ocean warms the ice, not only did the polar bear lose his iceberg home and the sea level rise to flood Mr. John's log cabin, but Mr. John's dog didn't have any more land to play on. And it was important that he had a name, and his name is Chico.

Here in this school tucked behind a fast-food restaurant on a main highway that whips cars to and from the Jersey shore, learning about the fascinating and interconnected systems of life on our planet is growing into every aspect of the day. Scavenger hunts for found items on the playground, songs sung at circle time, seed starting, and lunchtime sorting. Every day is a new discovery, and a testament to curiosity and the creativity that will evolve into caring and conscious stewards of our Earth.

Figure 4.1 A Preschool Class at Peppermint Tree Playing with Habitats in their Classroom. J. Hockenjos

While one child pockets an empty goldfish bag at the end of the day ("I am going to reuse this to put my beep-beeps in!"), another parent shares that her child wishes she could keep learning this in kindergarten next year. A mom shares her son's latest project. "He had stuck every drill bit and screwdriver he could find in the ground in our backyard," she said. "I asked what he was doing, and he told me they were 'cooling sticks.' They are sucking up the hot air in the earth and sending it back to the sun."

Bringing climate education into the classrooms of our youngest learners has the potential to strengthen our communities, enrich our schools, and reframe our outlook on a changing world. Whether it is one lesson a day or one lesson a week, children having the chance to fall in love with their natural and built environments and learn about climate change is catalyzing an anthropogenic force for good. We educators have the opportunity to build a foundation of environmental literacy and climate resilience not yet seen in modern history.

Notes

1. Madeline Will, "If Climate Change Education Matters, Why Don't All Teachers Teach It?" *Education Week*, March 3, 2023, https://www.edweek.org/teaching-learning/if-climate-change-education-matters-why-dont-all-teachers-teach-it/2 023/03

2. Lauren Madden, *Report on K–12 Climate Change Education Needs in New Jersey* (Trenton, NJ: New Jersey School Boards Association & Sustainable Jersey, February

2022), https://www.njsba.org/wp-content/uploads/2022/02/climate-change-ed-online-2-2.pdf

3. Alliance, Natural Start. "Nature-based Preschool Professional Practice Guidebook." (2019).

4. North American Association for Environmental Education. "Guidelines for Excellence: Environmental Education Materials. (2021)

5. Alicia Nortje, "Piaget's Stages: 4 Stages of Cognitive Development & Theory," Positive Psychology, May 3, 2021, https://positivepsychology.com/piaget-stages-theory/#operational

6. Italicized text represents language educators can use with students during the lesson.

CHAPTER 5

Climate Change Education in the Elementary Classroom

Jeanne Muzi

Considering the why, when, and how . . . and cultivating critical thinking and creative problem-solving while strengthening perspectives of resilience with young learners

When New Jersey became the first state in the nation to include standards to support climate change learning, it didn't take long for educators across the state to begin having some very difficult conversations about this approach to climate teaching and learning. Following those discussions, questions quickly emerged from colleagues, family members, neighbors, and the community. Upon analysis, it became clear that most questions fell into three different categories: Why are we introducing climate change to young children? When could learning about climate change possibly fit into the already-packed schedule of the average elementary school day? And, most importantly, how could these extremely complex concepts about climate change be presented to young students so they could be understood without being frightening and overwhelming? This chapter explores examples from one K–3 school in Central New Jersey using climate change education to spark problem-solving and critical thinking, from the perspective of the school's principal.

The "why" questions were actually the easiest to answer. As the devastating effects of global climate change continue to impact life on Earth, it is important to plant the seeds of problem-solving and resilience within our youngest students so they have ample opportunities to envision solutions, dream big ideas, and tap into all possibilities that may exist now and in the future. When schools can do that, our students can grow up with an appreciation of the world that surrounds them and an understanding of how we live with the results of our actions. They can develop critical thinking and look for meaningful solutions. They will not be stuck in the past, but instead seek out innovation and originality. The young students currently sitting in elementary schools around the globe are already experiencing many aspects of climate change. By providing a wide range of authentic learning experiences that teach about climate change through a lens of possibilities, we can equip all students with the knowledge, awareness, and understanding necessary to become informed, conscientious, and empowered stewards of the environment who are ready to solve some of the most challenging issues of climate change. The "why" is clear . . . we need problem solvers.

As six-year-old Aicha wrote, "It looks like all this harm is happening because people aren't working together. When you work alone, you sometimes don't care. Let's work together so we all care even more."

The "when" questions recognize that an elementary student's day is filled from the moment they arrive at school through dismissal. Teachers juggle math, science, reading, writing, spelling, art, music, physical education, character education, library, health, and more! Teachers recoil at the idea of "One More Thing!" Instead, meaningful and authentic interdisciplinary learning experiences needed to be developed that would weave several relevant learning standards together so students could excel in a cross-curricular manner while integrating higher-order thinking such as perspective-taking, concept synthesis, and justification of conclusions. For example, first-graders set up sunlight observation centers around their school campus. For several months, they collected data from the centers (temperature, shadow size, growth of plants, etc.), and then analyzed the data and drew conclusions. They determined how higher temperatures impacted their surroundings. The students then used a variety of materials to design and build a structure that reduced the warming effect of the sun on their observation centers. Every student was highly engaged throughout this unit and their questions and conclusions constantly fueled classroom discussions and reflections. They followed up the engineering experience by interviewing a climate scientist via Zoom, created a zine of their learnings and wonderings, supplemented the facts with poetry and illustrations about their new knowledge, and then presented their work in a performance to the community. The integrated approach made for a time-saving, thought-provoking, and memory-making experience for each of the first-graders. Utilizing a collaborative, action-oriented approach to asset-based learning, this unit placed the students at the center, driven by their own curiosity and passion, and introduced climate change concepts in an age-appropriate manner.

The song first-graders Miguel and Jon wrote included the lyrics, "I know we can fix the world. We can do it with our brains. With lots of new ideas, we can solve the earth's pains."

The "how" questions were the most difficult to explain. How we choose to teach young students about climate change is really the most important decision we make as educators when designing relevant, authentic climate learning experiences for our students. As adults, we recognize that climate change is the greatest and most pervasive threat to the natural environment the world has ever seen. We know all communities will be affected by climate change. But elementary school students should not be introduced to climate learning through a lens of risk and devastation. As an elementary educator, I know posing questions and guiding inquiry about climate requires a thoughtful and positive approach in order to inspire a spirit of hopefulness. Teachers know how to avoid overwhelming students with complex details, statistics, and concepts far above their developmental level. The goal is to always focus on conveying the importance of taking care of the planet in a way that makes sense to our students. We have found that young students find great comfort learning about the resilience of nature and how ecosystems can recover when given the chance. Teachers communicate that every positive action, no matter how small, contributes to building a more resilient and sustainable planet (this realization led to a group of second-graders championing the revitalization of a plastic bottle recycling program in the hallways of our school, which changed how single-use plastics were used throughout our building).

To commit to our objectives of cultivating problem-solving, critical thinking, and resilience, it is important to routinely bring awareness to students about successful initiatives and when people just like them made a difference. Every community has scientists, activists, or community members who have taken steps to protect the planet. Introducing these people to our students lets them see how small actions can lead to significant change. Letting students know about projects where people have made a difference in addressing environmental challenges, such as cleaning up a local river or enacting the Endangered Species Act, which has successfully protected wildlife, or how adopting a sustainable practice led to a change in behavior (such as the banning of plastic bags in New Jersey) gives students hope. Students need to understand that each of us have choices and our decisions demonstrate how we relate to our planet. Young students whose actions have fostered a sense of responsibility about their world grow to be citizens who embrace environmental stewardship and are more likely to seek solutions than simply accept the status quo.

Taking steps to solidify the why, when, and how of climate change education requires a school-wide collaborative effort. Slackwood Elementary School is a small, culturally rich K–3 school in Lawrenceville, New Jersey. One of the most difficult aspects of bringing climate learning to our students was determining where to start. So, as a community, we made three concrete decisions:

1. **Commit as a school:** While we recognized climate change education holds the promise of making school more meaningful and relevant for more students as it establishes connections to the real world and places them in it, we also knew as the educators charged with sharing these complex concepts that we were coming at it from a variety of entry points. Instead of just throwing

climate change immediately out to all the teachers, we decided the first step was to establish common understandings. Through a committee that included the school counselor, librarian, and enrichment specialist, a grade-by-grade framework was developed that would establish common definitions, shared understandings, and grade-level themes. The four themes would build upon each other. Each grade level included hands-on lessons and activities. We called the program Slackwood Climate Kids, and with support from our education foundation, we purchased books and supplies to get the program off the ground. Slackwood Climate Kids is now in its third year and will remain in place to establish common understandings for all children and to help bolster the prior knowledge many need to learn so they can contemplate climate change and its impact in their classrooms as they embrace authentic interdisciplinary learning experiences. At each level, consistent definitions for the terms *greenhouse effect, fossil fuels, renewable energy, recycling, sustainability,* and *conservation* are introduced. The difference between climate and weather is taught, and data literacy is stressed, so all students strengthen their abilities to communicate, understand, and explore ideas through data.

Kindergartners begin their Slackwood Climate Kids experience by examining plants and talking about what living things need to grow. The students roll a ball of yarn as each child takes on the role of a tree or animal, thus creating a thick web as they come to consider the interconnectedness of the world. But once a scissor starts cutting the strands of the web, they clearly see how fragile our world is and how we depend on each other. They also spend time looking at flowers. While taking apart the flowers, they learn how to identify pollen and experiment with moving pollen from one place to another. Then our youngest students learn about bees, food, the declining populations of pollinators, and new ideas in agricultural engineering. First-graders learn about how much garbage human beings create and what happens to all of it. The students find out about plastics and have some opportunities to repurpose common school items. They learn what biodegradable means and get their hands dirty making recycled paper. In second grade, students learn about clean water, threats to the ocean, and hydropower, and they experiment with the best ways to clean oil spills. They learn about how damaged ecosystems are restored and advances in technology that protect nature. Finally, in third grade, the Slackwood Climate Kids learn about how higher temperatures lead to melting ice, sea level rise, and changes in currents. They learn about storms and experiment with heating and freezing. The students have an opportunity to engineer models that can withstand storm surge and learn about some of today's innovators and their solutions for living with climate change.

Slackwood Climate Kids has led to a continuum of knowledge built upon consistent understandings. Teachers are now able to incorporate additional texts, projects, and learning experiences that integrate climate science in a more holistic way knowing the students have a common background.

2. **Commit to outdoor learning:** When students develop a personal connection with nature, they are instilled with not just a sense of ownership, but also an obligation for its well-being. Stepping outside into the natural world during the school day fosters a love for the environment. Research shows that students who have opportunities to participate in outdoor learning develop a positive sense of self, increased independence, confidence and creativity, improved decision-making and problem-solving skills, and greater empathy toward others. By participating in outdoor activities and exploration, our students became more invested in their immediate community, but also more interested in the role they play in our larger global neighborhood. Teachers created seasonal walking path activities and students had the chance to observe local ecosystems, identify plants and animals, and learn about the importance of preserving biodiversity. Themed walks were created, such as "Mindful Walks," "Collector's Walks," and "Data Walks." Students came up with ideas like "Story Walks" and "5 Things I Noticed Today Walks." They had the chance to participate in basic environmental monitoring projects as they tracked temperature changes, observed weather patterns, examined tree growth, and measured air quality. Through these ongoing efforts, students strengthened their abilities to examine data and analyze trends over time. Each season, we are reminded that well-designed and implemented outdoor learning experiences complement our curriculum and ignite a new sense of wonder for many students.

Through our teachers' efforts, students got involved in the clean-up day at a local lake, enjoyed planting trees out in the community, and shared the leafy green vegetables grown in their hydroponic garden. Through all these activities, students also learned about plant life cycles, soil health, and sustainable gardening practices. By discussing the impact of these actions on the local environment, our students have learned about the power of collective efforts, which is precisely the power necessary to solve the climate challenges we are facing. Being outside learners has led to storytelling, poetry writing, and watercolor painting sessions among the trees. From kindergarten through third grade, students learned about invasive and native plants. They learned to set up bird feeders and how to track woodland animals from the surrounding area. They have become stewards of their environment and are advocating for a cleaner, greener community.

One of the easiest and most effective ways to improve the quality of outdoor learning is to mount several observation checkpoints along where the students walk. These simple signs require each student to stop and observe. Sometimes the checkpoints are accompanied by a question. Once they read the sign, students can reflect silently, turn and talk to a friend, or write their ideas in a journal. Since the checkpoints remain in position all year long, each time the students stop to observe, they see something new as the seasons change or the weather surprises us.

Richard Louv, author and co-founder of the Children and Nature Network, once wrote, "How can our kids really understand the moral complexities of being alive if they are not allowed to engage in those complexities outdoors?"[1] Spending time in nature and becoming connected to the larger world is a game changer for many students. Once our school committed to outside learning, it became easier for teachers, and thus students, to become involved in community outreach projects related to environmental stewardship. Collaborating with local environmental organizations and community partners has meant our students have more opportunities to learn about the work being done locally and globally in the climate arena. Acting with intent to make the world a better place means growing a generation of citizens who see themselves as problem solvers and activists beyond the classroom.

As seven-year-old Sam concluded, "So that means your water is my water, and my water is yours."

3. **Commit to using climate learning as a lever to encourage higher-level thinking:** Learning about climate change involves understanding complex scientific concepts. Teaching children about climate change encourages the development of critical-thinking skills as they explore its causes,

Figure 5.1 Observation Checkpoint Signage created focused areas for students to notice and wonder. J. Muzi

consequences, and potential solutions. It promotes a scientific mindset and curiosity about the world. Educational experiences that integrate climate-change issues and challenges help develop students' abilities to thrive in the twenty-first century. It is imperative that students understand systems thinking, embrace collaboration, strengthen their questioning skills, hone their ability to communicate, seize opportunities for creative problem-solving, and sharpen their global perspective. The seeds for these high-level skills must be planted in elementary school and carefully tended throughout our students' time in school so they are well-prepared for the world that awaits them as adults. Bringing a "Thinking Set of Tools" to conversations and teachable moments regarding climate change prepares students to be lifelong learners. These tools include:

- Venn and tri-diagrams: The ability to compare and contrast focuses our students to draw conclusions and solidify their ideas. Young students who can compare animals, plants, ecosystems, storms, etc., become high-schoolers who can compare salinity of water, extinction rates, the effectiveness of wind power compared to solar, and so on. By continuing to build out a student's repertoire of graphic organizers with options like problem/solution charts, concept maps, and if/then diagrams, teachers can strengthen thinking and problem solving across content areas.
- HOTS/MOTS/LOTS: Students should be able to pose a variety of questions (higher-order thinking, medium-order thinking, and lower-order thinking) and figure out how to use the various questions to make sense of a dilemma or draw a conclusion. A teacher may ask for one of each level of question or have the students generate an array of questions and then categorize them. Learners improve as questioners when they are provided with lots of opportunities to practice questioning.
- Time for reflection: Climate change provides deep opportunities for students to reflect and ponder. This can be done in their journals, recorded on a tablet, or added to a collaborative message board. The ultimate intent of teaching reflection is for students to strengthen the habit of contemplating their own actions and constructing meaning from their own experiences. Reflection time is the perfect opportunity to consider what was learned and what still fills the student with curiosity.
- List activities: Strengthen students' abilities to evaluate what they know and prioritize one idea over another by using lists including "The Top Things I Know About," "The Three Most Important Things I Know About," "Two Things I Don't Want to Forget About," "Four Things I Still Wonder About," and so on. All these list activities provide time and space to ponder issues of climate change. We have had kindergarteners write the top things they know about polar bears and third-graders write about the three most important things they know about solar energy. Stu-

dent partners shared four things they were still wondering about pollinators and another group of students stated the two things they did not want to forget about fossil fuels and plastics. A list activity provides a quick start for all students, especially those who are thwarted by a blank page. The list activity gets kids thinking and talking to peers. It grows with students and is self-differentiating. It also helps students learn to justify their positions and make their case for something they feel strongly about.

- Wonder and notice: This is a familiar thinking tool that can be deployed into all lessons, conversations, and activities when centered on climate change. My students have used wonder and notice while watching videos of the ocean, reading articles in magazines, on walks in the woods, and during hands-on experiments. Wonders and notices can be done individually or with a partner or group. The act of observing, thinking, and questioning should be brought to the forefront of learning every day. Climate change can be an effective vehicle in developing these skills.
- The climate change book of the week highlights one text and provides time for a teacher read-aloud, question collection, and student reading time. As more age-appropriate and thought-provoking books are published for elementary students, creating a showcase for a single book communicates a school's commitment to climate learning and fits into a teacher's day. Using Post-Its and students' choice and voice in book selection, a single showcased book can open up time for meaningful discussion. Some books to consider include *Thank You, Earth: A Love Letter To Our Planet* by April Pulley Sayre; *Moth: An Evolution Story* by Isabel Thomas and illustrated by Daniel Egnéus; *Fatima's Great Outdoors* by Ambreen Tariq and illustrated by Stevie Lewis; *We Are Water Protectors* by Carole Lindstrom and illustrated by Michaela Goade; *The Boy Who Harnessed the Wind* by William Kamkwamba and Bryan Mealer, with pictures by Elizabeth Zunon; and the ever-popular, *Magic School Bus and the Climate Challenge* by Joanna Cole and Bruce Degen.
- Large-scale data collection: Young students learn about data in a multitude of ways in elementary school. From creating bar charts of the weather to graphing favorite foods and colors, early experiences with data encourage students to look deeper and draw conclusions. Once students begin making outdoor observations and gathering data over time, a highly engaging tool is a large-scale data-collection display that can be added to a hallway, classroom wall, or bulletin board. After some initial planning, ownership of the data should be handed over to the students. As you can see in figure 5.2, an example of a data-collection display made from two years of stream observations and information collection, the long view is encouraged. Photographs, diagrams, illustrations, pie charts, bar graphs, and more merged to tell the story of a stream environment and how the water, microorganisms, and plant life were changing right in our neighborhood.

Figure 5.2 A Collaborative Data Collection Display helped students curate the information they gathered over a school year which guided them to make clear connections and draw reasonable conclusion in a more accessible way. J. Muzi A Collaborative Data Collection Display helped students curate the information they gathered over a school year which guided them to make clear connections and draw reasonable conclusion in a more accessible way. J. Muzi

When young students gather data about the environment and display the information, they can gain valuable educational experiences and contribute to a broader understanding of environmental issues. A large-scale data-collection project can strengthen and develop critical-thinking skills and help students learn to interpret data, logically draw conclusions, and make informed decisions based on evidence.

When this collection began, the students were first-graders, and the project continued into their second-grade year. With experience, the students learned how to formulate hypotheses and draw conclusions based on evidence. For example, they came to understand that the runoff from a nearby golf course was contributing to the change in water quality. Through debate and perspective taking, the students took both sides of the argument and learned that "right answers" are often not so cut and dry when examining complex environmental issues.

Every month, the students gained practical experience and learned about the importance of accurate and reliable data through gathering measurements and seeing the results of experiments. Since the students were highly invested in sharing information with their peers, they worked hard to identify long-term patterns and trends in the data. They learned to organize information in a clear and coherent manner. Most importantly, they learned the value of collaboration and learned that any of the complex issues facing the world today requires different people with different experiences to come together to share their knowledge, ideas, and strong work ethic. This highly engaging endeavor can become a project that encompasses scientific skills, critical thinking, communication, teamwork, and a sense of environmental stewardship. As the students get older, climate change learning will involve analysis of vast amounts of data, including temperature records, carbon dioxide levels, and ecosystem changes. Planting the seeds of data literacy early makes it possible for students to learn to question, analyze, and synthesize information, which are essential skills for addressing the challenges of the modern world.

As six-year-old Leanne presented the data-collection project, she said, "Once we looked at everything together, we saw how it was all connected, and we knew how we could protect our stream."

> By filling our students' "Thinking Toolboxes" with tools that can be utilized across all content areas, they are able to understand how multifaceted issues of climate change truly are. As they begin to see the interconnections between people, places, and things, children can begin to pull strands together from science, social studies, geography, math, and civics and consider how human activities impact the environment and what the consequences are. Committing to conversations and readings, along with questioning and observations that focus on climate change, encourages students to think creatively about possible solutions and consider the long-term impacts of different approaches.

Elementary educators recognize the enormous responsibilities they possess since all future learning must be built on a very strong foundation. They deliberate and ponder every decision they make to ensure every moment of learning is designed to support their students across all subjects. Climate change learning must not only fit into the elementary day, but it must also provide powerful learning outcomes to deserve precious instructional minutes of the elementary day. The integrated approach and cross-curricular thinking focus provides for that.

Eighteen-year-old Logan wrote on his college application, "I became an environmental scientist when I was six years old and my teacher took me and my friends out of the classroom and into the woods. That changed everything for me."

There is no question that teaching about climate change in elementary school is challenging. We are aware of things like melting glaciers, rising sea levels, extreme weather events, and changes in various ecosystems. Even so, we must share some of the good news with our students: Global climate activism is at an all-time high; energy from wind, the sun, and water continue to grow in popularity and cost-effectiveness; and, according to the National Oceanic and Atmospheric Administration, "Over time, steady progress is being made and the hole in the Ozone Layer is getting smaller."[2] Our students live in historic times. The consequences of climate change are not off in some distant future; they are being experienced right now. It is the reality that today's children have inherited and what they will need to navigate for decades to come, which pushes the urgency of focused climate learning to the forefront.

When elementary teachers ensure all climate-change learning is rooted in facts and current science, they can arm their students with knowledge that enables them to speak with understanding and conviction. Educating children about climate change empowers them to make informed decisions in the future. It instills a sense of agency, encouraging them to think about their actions and the potential consequences of those actions. When students understand the consequences of human actions on the environment, they are more likely to adopt sustainable practices and advocate for positive change. This sense of responsibility lays the groundwork for cultivating environmentally conscious citizens who are committed to preserving the earth for future generations.

Our students must understand they do not need to go alone. Inspired and engaged groups of people emerge every day to come together and figure out new solutions and ideas to mitigate the impact of climate change. We need to remember there are many positive developments in the climate change arena and we must convey this spirit of hope to our students.

Through an integrated approach, educators in New Jersey have the opportunity to prepare students to understand how and why climate change is occurring and learn how to share the impact it has on our local and global communities. Through a system of common understandings across a school, a commitment to outdoor learning time, and a toolbox full of thinking tools, elementary educators can inspire, engage, and guide students so they can come of age acting in informed and sustainable ways.

The science is clear: The earth is undergoing unprecedented changes due to greenhouse gas emissions, deforestation, and the unsustainable exploitation of natural resources, which has resulted in more frequent and severe weather events, rising sea levels, loss of biodiversity, and disruptions to ecosystems. The scope of the problem solving our students will be required to do is daunting, which is why the work of elementary educators is so important, as we balance facts with hope, reality with possibilities, and rigidity with adaptability and innovation. Once educators who work with young children see the importance of the why, when, and how, we can connect the here and now with the future.

Our school motto at Slackwood Elementary School is "Together we can." Through that phrase, we inspire our students to see that collectively, we are so much more powerful and impactful than we could ever be as individuals. We consciously communicate to our young students that the path we must travel to mitigate climate change has room for all interested citizens. They need to know every person can step into the role of climate problem solver. This realization is key to having students discover their own power . . . creatively, intellectually, and humanistically. Together we can teach climate change well and strengthen all students' abilities to be part of the solution.

Notes

1. Richard Louve. *Last Child in the Woods: Saving Our Children from Nature-Deficit Disorder* (Chapel Hill, NC: Algonquin Books of Chapel Hill, 2008).

2. Simon Torkington. The ozone layer is on the right path to recovery: Here's how the world made it happen. World Economic Forum. (2023). https://www.weforum.org/agenda/2023/09/ozone-layer-hole-update-nasa/

CHAPTER 6

Education for Climate Action in Secondary Schools

Andrea Drewes and Jessica Monaghan

Climate Change Education

Climate change education in middle and high schools is essential for fostering environmental literacy, critical thinking, and active citizenship. Recent efforts to incorporate climate change into curricula have gained momentum to equip students with the knowledge and skills they need to understand, address, and mitigate climate change impacts on the local and global scales. Traditionally taught in science classrooms, climate change education includes understanding the phenomenon's causes, impacts, and potential solutions. This approach emphasizes scientific concepts like the greenhouse effect, carbon cycling, and the role of human activities in driving climate change.[1] In secondary science classrooms, students analyze data, develop explanations, evaluate policy, and propose mitigation strategies at various levels of complexity.

Climate change discussions in science classrooms can also extend beyond science concepts to explore ethical considerations and societal implications. Science teachers may incorporate the role of governments, industries, and individuals in shaping climate policies. By fostering an understanding of the interconnectedness across science, ethics, and policy, students are equipped to become informed and engaged citizens who can advocate for positive change.[2]

However, research highlights the limitations of teaching climate change solely within natural sciences, documenting that many teachers have a narrow view of climate change and overlook its broader interdisciplinary nature.[3] Addressing climate change effectively requires insights from diverse fields, including social sciences, health sciences, politics, ethics, and the humanities. Current efforts to integrate climate change education across all disciplines are progressing. New Jersey is the first (and currently only) state to require climate change education across all subject areas and grade levels. The New Jersey standards seek to "foster an interdisciplinary approach to climate change education that is evidence-based, action-oriented and inclusive."[4] Other states like Connecticut, California, and Maine are making similar efforts but have not yet mandated interdisciplinary climate change education.

In the humanities, students may examine environmental movements' historical context, analyze the geopolitical implications of climate change, and explore climate justice issues. Social studies curricula can offer a sociocultural perspective on climate change, enhancing students' understanding and appreciation of its societal impacts. Language arts classes may focus on climate change literature, multimedia analysis, and persuasive communication to help students develop empathy and advocacy for environmental causes.

Efforts to enhance climate change education in secondary schools should focus on fostering critical thinking skills, promoting active learning experiences, and empowering students to become agents of change in addressing the climate crisis. In addition to studying the phenomenon itself, research recommends climate change should be viewed as more than just a scientific issue to enable students to develop critical thinking skills[5] and more fully consider solutions to climate change.[6] Similarly, Allen and Crowley[7] argue that increasing climate science knowledge alone is not sufficient for effective climate education. They emphasize the importance of increasing collective efficacy through participation, relevance, and interconnectedness in learning experiences to facilitate meaningful responses to climate change.

Bicycle Model: A Metaphor for Climate Change Education

The bicycle model metaphor for climate change education Cantell and colleagues present[8] offers a comprehensive framework for understanding the multifaceted nature of addressing environmental issues. By likening the process of combating climate change through education to riding a bicycle, the metaphor emphasizes the interconnectedness of individual actions and collective decision-making in progressing toward sustainability. Just as a bicycle requires effort and coordination to stay balanced while moving forward, addressing climate change also requires a concerted effort from individuals, communities, and societies to make meaningful progress.

Within this metaphor, each bicycle component represents a key aspect of climate change education. The interconnected nature of the *wheels* underscores that students must understand the concepts and impacts of climate change while refining critical thinking skills to evaluate information and envision alternative scenarios. While knowledge is valuable, it is not sufficient for people to undertake sustainable action on climate change.⁹ Knowledge, action, participation, and other elements must be supported by the interpretative framework of our personal and societal values, identity, and worldviews, much like the bicycle's *frame*.

The *pedals* symbolize individual actions that contribute to mitigating climate change, such as reducing carbon emissions and sustainable practices. Taking climate action and moving forward on a bicycle both require some effort to be successful. The bicycle's *seat* makes the effort to pedal much easier. In this metaphor, the seat also demonstrates the need for motivation and participation with like-minded, supportive individuals to make collective progress on climate related initiatives. The handlebars and brakes control the direction and speed at which the bicycle can move. The *handlebars* indicate the forward-thinking direction climate change education should maintain as it steers toward possible solutions and action. Meanwhile, operational barriers arise, much like *brakes*, which cause delays and limits in progressing toward climate action. Finally, the model also notes the role of emotions in shaping individuals' attitudes toward environmental action, highlighting the need to consider ethical issues and cultivate realistic hope for the future. These emotions, especially hope, are represented by the bicycle's *light* shining a path forward.

By incorporating these elements into climate change education, the bicycle model serves as a valuable tool for guiding educators in fostering a comprehen-

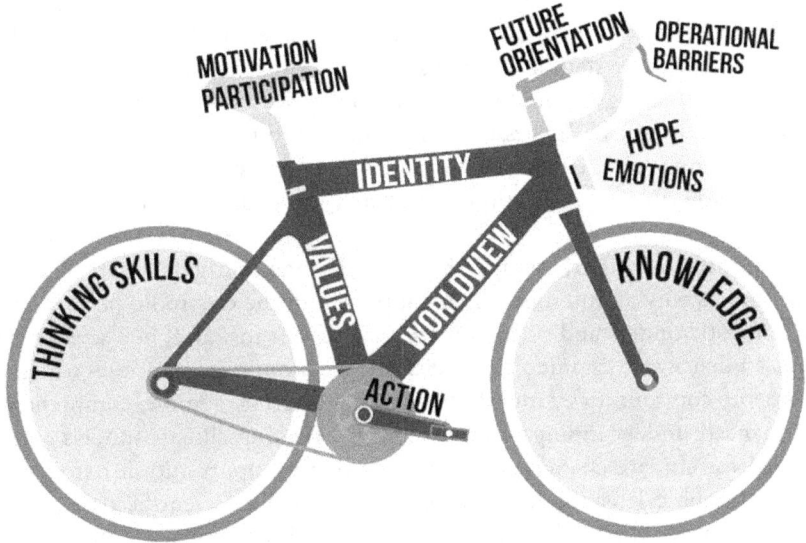

Figure 6.1 Bicycle Model. Image from Cantrell et al., 2019

sive approach to inspiring and empowering students to take meaningful action. In the following sections, we will break down this model further and align the components to best practices for climate change education and explore examples of these methods through vignettes in secondary-school settings.

Moving Forward with Interdisciplinary Connections: Wheels of Knowledge and Skills

Successful climate change education requires a shift in responsibility from full reliance on secondary science classrooms to a multidisciplinary approach. Effective climate change education necessitates a blend of content knowledge and critical thinking skills for students to comprehend the multifaceted nature of this global issue. Through an interdisciplinary approach, students can make connections between science, social studies, language arts, mathematics, and even foreign languages to appreciate the complexities of addressing this pressing issue.

Cantell et al.'s[10] bicycle model emphasizes the essential role of interdisciplinary collaboration in climate change education. Just as a bicycle needs both wheels to balance and move forward, climate change education relies on the interaction between content knowledge and critical thinking skills. Through an interdisciplinary lens, students can explore various aspects of climate change, including natural phenomena and human impacts, fostering a holistic understanding of the environmental system.

Content knowledge development serves as the starting point for interdisciplinary climate change education at the middle and high school levels. A comprehensive understanding of the scientific principles underpinning climate change, such as the greenhouse effect, carbon cycle, and human impact on the environment, is crucial. It is essential for students to recognize the environment as a system and such the feedback loops that occur within it; to describe climate effects and the impact mitigation and adaptation strategies can have on those effects.[11]

The interdisciplinary nature of climate change as a global and local issue demonstrates why an interdisciplinary approach in the classroom helps students gain a holistic understanding of climate change. For instance, in a science classroom, students can examine the social and economic implications of climate change on communities around the world, fostering a more comprehensive and nuanced understanding of the issue. Investigating climate models and the evidence for climate change provides an in-depth opportunity for students to interpret graphical representations, collect and analyze scientific data, and explore mathematical models of climate change. This approach to climate change education integrates both math and science practices found in the New Jersey

Student Learning Standards (NJSLS-Science and NJSLS-Math) and in national Next Generation Science Standards and Mathematics Common Core standards documents.

In social studies, students can explore primary and secondary sources to identify historical moments that led to climate change contribution through industrialization, which is an important concept found in the C3 Framework for Social Studies and NJSLS-Social Studies. Students can consider the implications of climate change regionally by noticing local and global trends, impacts on movement of people, resource availability, and so on. Discussing climate action on local, state, national, and global levels helps students develop civic engagement and active civic participation skills.

Literacy and scientific literacy underpin all this work. Reading complex texts and written communication are essential for students to gather evidence about climate change and express their own ideas. Speaking and listening are critical skills for engaging with diverse perspectives and resources on climate change that might be represented in oral history traditions, science communication, and public speeches. Students can investigate the habits of effective communicators by analyzing primary documents and using multiple modalities to express their knowledge, both key outcomes included in the NJSLS-ELA and in the ELA Common Core standards.

Critical thinking skills paired with knowledge development enable students to analyze information, evaluate different perspectives, and draw well-reasoned conclusions. By fostering critical thinking skills, educators can help students question assumptions and misinformation, consider alternative viewpoints, and make informed decisions about environmental issues. Sezen-Barrie et al.[12] highlighted the importance of historical climate data interpretation in fostering critical questioning and understanding of climate mechanisms and human impacts. Critical thinking alongside systems-thinking enables students to engage in meaningful discussions and debates about climate change, encouraging them to think deeply about the problems root causes and potential solutions.

Integrating content knowledge and critical thinking skills empowers students to take action in their communities. For example, students (and teachers) who possess a strong foundation of content knowledge about climate change are better equipped to critically evaluate the sources of information they encounter and distinguish between reliable and unreliable sources.[13] Teaching students to assess information critically helps them become savvy consumers of environmental information and advocate for evidence-based solutions to climate change. These skills are essential in the digital age, where misinformation about climate change is still widespread.

The development of critical thinking and problem-solving skills can be infused into classroom learning experiences through climate change education. Basche et al.[14] highlight the effectiveness of problem-based learning (PBL) in promoting critical inquiry and student autonomy. By integrating local environmental issues and encouraging self-directed inquiry, PBL empowers students to construct informed arguments and engage in political discourse related to climate change. This interdisciplinary pedagogical approach enhances students' science attitudes and engagement and encourages synthesis of classroom learning with real-world applications for climate action.

When teachers engage with meaningful partnerships beyond the classroom, they can provide students with opportunities to apply their knowledge in real-world settings, connect with experts in the field, and contribute to environmental initiatives in their communities. Collaborating with experts enriches students' climate change education experiences and fosters collective efforts in addressing environmental challenges. Engagement with external organizations, local communities, and environmental advocacy groups can further enhance students' understanding of climate change and empower them to take active roles in environmental conservation efforts.[15] These partnerships inspire students to become informed and engaged citizens who address climate change.

Now we share high school and middle school examples of how New Jersey educators brought this interdisciplinary and collaborative framework to their climate change educational endeavors. These vignettes highlight the interwoven nature of the bicycle's wheels for content knowledge and thinking skills development in climate change education.

HIGH SCHOOL VIGNETTE: CLIMATE CHANGE PROJECT

Teachers implemented an interdisciplinary PBL unit in an urban public high school with a large percentage of bilingual students. The school had been newly established as part of a state initiative with a cohort model of twenty-five to thirty ninth-grade students. Students began the Climate Change Project with a kickoff event linking multiple disciplinary areas into the problem introduction. Students rotated through stations that integrated critical thinking skills across disciplines to show climate change's multifaceted nature. In one interdisciplinary station example, students created six-word stories to explain carbon emissions, sea level rise, and global average temperature graphs, thereby merging math, science, and literacy.

After the project's kickoff, students selected teams to tackle the problem scenario, choosing from interdisciplinary projects like designing a sustainable structure, developing a product to reduce climate change, changing a policy in the school district to reduce its carbon footprint, or giving a speech about climate change to the local community. Students worked on their projects during advisory time, while teachers integrated climate change concepts into their lessons. In this model, students progressed in both skill development and content knowledge.

Through professional development, teachers identified a collaborative timeline for implementing interdisciplinary PBL, addressing standards in mathematics, English as a second language, social studies, science, engineering, world languages, and ELA. A special educator was involved in planning and teaching across disciplines. Based on the scope and sequence of curricula, teachers mapped out essential questions, standards, and formative assessments to deepen interdisciplinary connections for teaching climate change, as seen in Table 6.1. In their social studies classes, students explored maps to identify people's migration patterns in response to sea level rise. In ELA, students researched passive and active efforts in response to climate change, then reflected on their role in contributing to or preventing climate change in their daily lives. While the day-to-day schedule and classroom structures remained, teachers learned from each other's content area expertise. The interdisciplinary map was a useful process and reference for teachers' planning, as demonstrated in Table 6.1.

The interdisciplinary approach fostered collaboration among teachers and students. Teachers gained insights from each other's areas of expertise, while students developed autonomy in solving climate-related problems. The interdisciplinary PBL approach required skills such as critical thinking, reading, citing evidence, writing, creativity, problem-solving, visual design, and public speaking. For example, one group proposed a vegetarian and vegan fast food franchise to reduce greenhouse gases, which showcased their understanding of alignment between cause and effect. Students explained how the solution addressed the problem by reducing greenhouse gas emissions. While the solution effectiveness and depth of science knowledge varied across groups, student work demonstrated interdisciplinary learning, critical thinking, and problem-solving. Ideally, a team-based school approach where teachers across disciplines share the same group of students and have time to meet would be most successful for replicating this project. An adapted version of this project could be accomplished by streamlining interdisciplinary concepts to fit the context of their individual content area classrooms.

MIDDLE-SCHOOL VIGNETTE: HUMAN IMPACT PROJECT

A middle school setting is ideal for an interdisciplinary unit when the school structure includes a team-based model. In this vignette, however, students were not in a setting with a team-based model, so the unit's interdisciplinary nature occurred within the context of a sixth-grade science classroom. Literacy and mathematics standards in the NGSS[16] and resources from the United Nations' sustainable development goals[17] (SDGs) were integrated as guides for bringing

Table 6.1 Interdisciplinary Standards Alignment Map

Essential Question	Standard	Formative Assessments
In what ways are we passive and active toward climate change and how can we effectively combat climate change?	ENG. NJSLSA.W9. Draw evidence from literary or informational texts to support analysis, reflection, and research.	Students synthesize research to demonstrate ways they contribute or prevent climate change actively or passively in their day to day lives
How are humans affecting biodiversity? How does biodiversity affect humans?	Biology HS-LS4-5. Evaluate the evidence supporting claims that changes in environmental conditions may result in: (1) increases in the number of individuals of some species, (2) the emergence of new species over time, and (3) the extinction of other species.	Create a model that explains how climate change affects biodiversity.
How can climate change affect the maps and geography that we have learned in school?	SOC. 9-12. CS.3 [Skill]. Students use a variety of maps and documents to interpret human movement, including major patterns of domestic and international migration, changing environmental preferences and settlement patterns, the frictions that develop between population groups, and the diffusion of ideas, technological innovations, and goods.	Students will interpret maps that analyze the rise of sea levels as evidence of climate change as well as maps that trace the migration of population groups due to the climate crisis.
How are Global Carbon Emissions growing exponentially? How can one differentiate an exponential model from a linear model on the global carbon emissions?	MATH-ALGEBRA Linear and Exponential Models F-LE A. Construct and compare linear and exponential models and solve problems 1. Distinguish between situations that can be modeled with linear functions and with exponential functions.	Students will compare how a linear or exponential function could represent the model of Global Carbon Emissions.
How do you determine how to communicate important ideas to the community of New Brunswick using a poster or sign at a protest?	SPANISH WL: 7.1.NM.A5. Demonstrate comprehension of simple, oral and written messages using age- and level-appropriate classroom and cultural activities.	Students will create climate protest signs or posters against climate change in Spanish and English

in multiple discipline areas. Additionally, this example included a partnership with a local watershed to enhance the unit with experiential learning on topics of sustainability. The Cantell et al. bicycle model[18] could be applied more broadly to address teaching sustainability and environmental issues, as noted here.

The goal of the Human Impact Project was for students to understand humans' impact on the world around them as it related to one sustainability issue. The options were to investigate the role of energy use (in relation to climate change and renewables vs. non-renewables), freshwater availability and access, and material waste and consumption. The bicycle model metaphor also aligns with the local-global connection to sustainability. The project unit began with students collecting and analyzing data related to energy use, waste production, and water usage in their school. Students estimated gallons of water used daily, wasted materials at breakfast and lunch, and energy usage. They collected observational data and compiled additional information (e.g., number of students, average energy consumption of lights, and gallons per toilet flush) to estimate and calculate human impact during the school day. They projected weekly, monthly, and annual data. Students then used this information to contextualize the local problem scenario aligned with relevant global issues. They used literacy skills to explain the connection between local phenomena and a global problem.

The teacher partnered with a local watershed that had a LEED-certified building to provide real-world examples of current sustainability and conservation technologies and methods. Specifically, they learned how the design and construction of the LEED-certified building intentionally reduced human environmental impact. The partnership included an in-person assembly with a representative from the watershed facilitating a virtual field trip to the site. The event served as a grounding shared experience for students to learn key concepts in building design, water conservation, and hydrology. Additionally, a smaller group of students attended an in-person field trip to gather more hands-on experience across the three sustainability issues. They then turnkeyed their learning with peers.

At the end of the unit, students developed 3D models, posters, or written proposals to improve sustainability practices at their school. They engaged in collaboration, critical thinking, creativity, problem-solving, and mathematical reasoning, and they reported their increased ability to collaborate with peers, communicate with professional adults, evaluate evidence for quality, and create models to support their solutions. The science content knowledge they demonstrated in their projects required them to explain the problem and solution, evaluate existing technologies, and identify connections between local and global issues. Overall, students demonstrated their understanding that human activities contribute to global environmental issues and that multiple solutions can address problems. They also noted that solutions for one issue may contribute to other problems. By evaluating individual contributions to solving environmental issues such as climate change, students could connect their home and school lives.

Framing Learning and Motivation through Learner Knowledge: Leveraging Project-Based Learning

Learners "frame" their learning through cultural values, linguistic assets, and life experience. In the Cantell et al.[19] bicycle model, the frame is built upon learners' identity, values, and worldview, while the frame connects knowledge and skills (the wheels). The bicycle frame categories play a pivotal role in motivating students. If teachers do not spend time learning and tapping into student perspectives, challenges will occur at the stage of initial motivation.

Reflecting on human beings' roles in climate change creates personal connections to the issue with a sense of accountability. To boost motivation, climate change education must emphasize people's role in building and changing society.[20] Building relevance for learners is key to facilitating their sustained motivation, and utilizing school settings strengthens communities and creates positive engagement.

Students' perspectives frame how they think about and engage with content knowledge.[21] For students to build relevant knowledge and skills, the learning experience must leverage their assets. When prioritizing frameworks such as culturally responsive teaching (CRT),[22] differentiation,[23] and universal design for learning (UDL),[24] a teacher must know their students deeply and integrate this knowledge into planning and facilitating meaningful learning experiences. UDL operates on the assumption that teachers can anticipate learner variability in classrooms due to variations among humans.[25] Utilizing the UDL framework benefits learners with disabilities and addresses student variability based on age, gender, cultural background, and first language.[26] CRT acknowledges and values students' diverse cultural backgrounds and experiences with the intent to create inclusive and affirming learning environments. Additionally, according to Hammond,[27] CRT empowers students to move toward independent learning, which aligns with learner-centered approaches.

Teachers must consider students' identities, prior knowledge, belief systems, learning preferences, and more[28] to create learning experiences that center them. Learner-centered approaches such as PBL naturally embed practices from the above frameworks because their design allows the learner to bring in their own beliefs, skills, knowledge, and attitudes to the experience.[29] PBL has been shown to improve student motivation, collaboration, and problem-solving skills and helps them develop a deep and flexible knowledge base.[30] These skills are essential for students to engage deeply with climate change topics and move toward an action-oriented mindset. Framing climate change education through PBL experiences allows students to explore authentic, relevant topics, then demonstrate scientific engagement and learning publicly through products or performance.[31]

PBL begins with students defining an ill-structured, complex issue or topic with their teacher's guidance; frequently, they are authentic and meaningful topics to students.[32] Students collaborate with peers to develop solutions based on their exploration of different resources and multiple perspectives and investigations of data, while teachers support students by facilitating understanding through content and skill development.[33] Students' final products and performances demonstrate critical thinking and problem-solving, illustrate their values and real-world connections, and propose action. Basche and colleagues[34] found that using a PBL approach in a middle school science classroom boosted students' engagement, confidence, and understanding of environmental topics.

If educators do not prioritize motivational factors in their design and implementation of climate change learning experiences, students will be challenged to find meaning and connection to their identities, prior knowledge, and experiences. They will encounter operational barriers or "brakes," which hinder the bicycle's forward motion. Environmental responsibility can be hindered at multiple points, beginning with a lack of knowledge and progressing to a lack of action. When educators do not orient a learning experience to their learners, operational barriers appear in the early stages of learning, which prevents skill and knowledge building.

The high school Climate Change Project and middle school Human Impact Project examples illustrated below will build upon the context in the previous vignettes to demonstrate how learner knowledge was used to boost motivation. Specifically, they identify how connections to student identity, values, worldview, motivation, and participation in the projects' design helped anticipate and avoid operational barriers at the knowledge development stage.

HIGH SCHOOL VIGNETTE: CLIMATE CHANGE PROJECT

The interdisciplinary Climate Change Project tapped into students' identities, assets, and values to initiate and sustain their motivation for learning. Teachers anticipated learner variability, linguistic differences, cultural assets, and operational barriers in the project design. One intention of the project was for teachers to facilitate their learners' empowerment through choice. Teachers anticipated operational barriers by creating translated resources for emergent bilingual students, using technology, and obtaining support from ESL teachers.

The project tapped into student interest initially with a survey to determine the global issues they were most concerned about from the UN's SDGs.[35] They were asked to complete the following statement: "If I could pick the top 3 issues in the world that concern me most, I would work towards solving. . . ." Most students selected Climate Action, Life Below Water, and Clean Water/Sanitation. They were shown the survey results to introduce the topic of climate change as the first one to tackle, noting that teachers would address other topics in the future. Starting with student interest was a way to identify the topics students valued and felt passionate about solving. The teachers' transparency in sharing the survey results boosted students' motivation because they knew they were being heard.

The PBL unit began with the kickoff event described previously. The activities were designed with UDL in mind by including multiple means of representation, engagement, and expression, with specific ties to CRT.[36] Both UDL and CRT require teachers to be aware of their students' individuality, then develop supports based on learners' varied strengths, abilities, cultures, skills, and preferences.[37] Specific examples in the Climate Change Project kickoff event included watching a Greta Thunberg speech to empower students as youth advocates and highlight her multilingual identity, which resonated with many students. Highlighting speech and oral storytelling as an important form of communication honors a variety of cultural connections while providing multiple means of engagement and expression.

During the kickoff, students identified their favorite meals and reflected on the greenhouse gas emissions produced by making them; then, they were challenged to recreate the meal in a way that produced fewer greenhouse gasses. The focus on food tapped into cultural relevance for students, as they noted their commonalities and differences with peers and the options provided. They also had the opportunity to discuss and reflect on their perspectives as they built community with peers.

Additionally, PBL provides students with significant choices to sustain their motivation and engagement. Students selected group members with whom to collaborate; identified which aspect of climate change they would address (e.g., carbon emissions, local effects of climate change, communication with the public); and chose the type of product or performance to communicate their understanding. These aspects of choice align well with UDL principles to provide students with agency and ownership of their learning throughout the process. Additionally, the project helped build student capacity to self-regulate with varied support from teachers and peers.

MIDDLE SCHOOL VIGNETTE: HUMAN IMPACT PROJECT

In the middle school project, learning was contextualized within local phenomena to create relevance and motivation for students. Beginning with the school data collection, students were empowered to define the problem as a class and generate ideas about what concerned them most in their observations of building operations. While a multitude of perspectives, identities, and values were present in the discussion, classes reached consensus about people's roles in the school and their contribution to globalized issues.

Throughout the Human Impact Project, students were provided with mini-lessons to deepen their understanding of their individual problem scenarios and how they related to what their peers were exploring. For example, during a station rotation activity, students took apart an outdoor solar lamp to explore and explain the energy transformations that occurred inside it. They used this hands-on experience to understand how solar panels work and impact the school's energy bill. Multiple means of engagement allowed students to gain perspectives on the scale of renewables needed to avoid traditional power sources.

The watershed also provided highly novel experiential learning to motivate students. On the field trip, students played games to understand resource availability. They explored features of the LEED-certified building that related specifically to their own projects while learning about other student groups' issues. The experience provided current solutions and technologies in an observable, tangible context to connect observable local issues with abstract globalized issues.

Students also chose their project teams, resources, and means of expression for their final products. Project requirements included creativity and justification for proposed solutions. Students could not simply propose technology and solutions that already existed; they had to build upon concepts and solutions to create innovative solutions. With creativity, students included their identities and interests in solutions. For example, one group noted that the afterschool STEAM and art clubs could benefit from making art with recycled materials. Another group proposed installing pedals under desks to generate mechanical energy and help regulate "fidgety" bodies with movement while maintaining attention in the classroom.

Students evaluated the effectiveness of their potential solutions based on evidence from research, experimentation, or mathematical models. Based on student strengths and perspectives, they justified their solutions with varying methodologies and depths. One group conducted an experiment in the cafeteria to see if students would recycle and compost appropriately if a new policy was implemented. Another group analyzed the number and type of solar panels needed to generate the energy needed to balance the school's energy bill. Operational barriers were most visible in the requirements for creativity and justification because of the abstract thinking and mathematical skills required, so teachers accommodated or modified expectations as needed in the inclusion classroom. While these were unanticipated challenges, they were manageable. All students were engaged with creative, abstract thinking in their proposed solutions to varying levels of complexity.

The Power of Education for Climate Action: Overcoming Barriers to Acting through Collaborative Participation

Today, the urgency of addressing climate change is more critical than ever. The need for mitigation and building a sustainable future for future generations is increasingly evident. As highlighted in earlier sections, education is a powerful tool in the battle against climate change. An education for action (EfA) model with a solutions-oriented perspective can empower students in middle and high schools to become informed, engaged, and proactive agents of change.

EfA is a pedagogical approach that emphasizes empowering students to become active participants in addressing real-world challenges and move beyond passive learning. By incorporating elements of activism, advocacy, and social change into the educational experience, this model encourages students to apply their knowledge and skills to effect positive change in their communities and beyond.[38] In the context of climate change education, this approach enables students to understand the science and impacts of climate change and explore and develop solutions that contribute to sustainability and resilience.

One key benefit of using an education for climate action (EfCA) learning model is its ability to contextualize abstract concepts and scientific data in a way that is relatable and relevant to students' lives. By connecting climate change to local and global issues, students can see how their actions, choices, and behaviors contribute to the problem and, more importantly, how they can be part of the solution. This approach fosters a sense of responsibility, agency, and empowerment among students, motivating them to take action.

Cantell et al.'s bicycle model[39] highlights the importance of practical action in climate change education, emphasizing the need for learners to engage in real-life issues to effectively participate in climate change mitigation. Action is depicted as effortful, requiring motivation, empathy, and societal norms. This aligns with the notion that environmental action is influenced not only by knowledge, but also by emotions and the ease of acting.

Despite the growing awareness of climate change and its impacts, several barriers hinder meaningful action, symbolized by the bicycle's brakes. One of the primary barriers is a lack of urgency or perceived relevance among individuals who may not fully grasp the immediate threats posed by climate change. This disconnect between knowledge and action can be addressed through education that emphasizes climate change's real-world and local impacts and inspires a sense of moral responsibility to act now.

Another barrier to climate action is the complexity and scope of the issue itself. Climate change is a multifaceted challenge that requires interdisciplinary solutions and collective efforts on a global scale. This complicated nature can further slow students' ability to act, even in local settings. Permissions, funding, and resources can be challenging for student-initiated action projects. However, an EfCA approach can help break down this complexity by providing students with a comprehensive understanding of the interconnected nature of climate change and the various factors that contribute to its exacerbation. Students with the necessary knowledge and skills fostered by engaging with climate change education can overcome this barrier and engage in effective, student-driven action.

One final logistical barrier that secondary teachers often experience is a scarcity of time and the curricular resources this pedagogical approach requires. Teachers report feeling overwhelmed and unprepared to integrate climate change into their non-science classrooms when their standard curriculum expectations are already so plentiful. Teachers should envision themselves as guides to facilitate student action projects and persevere through challenges in an EfCA model. Central to the EfCA model is a solutions-oriented perspective that focuses on identifying and implementing practical strategies for addressing climate change. Instead of dwelling on the doom-and-gloom narratives often associated with the climate crisis, educators can emphasize the potential for positive change and opportunities for innovation, collaboration, and sustainable development.[40] The future orientation aspect that Sezen-Barrie et al.[41] discuss underscores the importance of envisioning the future critically but optimistically. Decision-making and creative thinking are crucial for shaping a positive future amid the complexities of climate change scenarios. By highlighting success stories, best practices, and innovative solutions from around the world, students are encouraged to think creatively, critically, and optimistically about the planet's future.

A solutions-oriented perspective also helps counter feelings of helplessness, apathy, or despair that can sometimes arise when confronting the magnitude of the climate crisis.[42] By demonstrating that individual and collective actions can make a difference, educators empower students to believe in their capacity to effect change and inspire others to join them in the fight against climate change.[43] This shift in mindset from problem-focused to solution-focused not only motivates students to act but also cultivates a sense of hope, agency, and resilience in the face of adversity.

Emotions play a significant role in shaping how individuals perceive, process, and respond to information about climate change.[44] EFCA recognizes the importance of addressing emotions in climate change learning by creating a safe space for students to express their feelings, concerns, and hopes openly. Hope and compassion are essential in stimulating positive attitudes about climate change, enabling individuals to engage in proactive solutions to address this pressing issue. Cantell et al.[45] describe emotions, especially the feeling of hope, as the bicycle's light, which serves to shine a path forward and diminish the gloom of negative emotions. By acknowledging and validating all emotions, educators can help students develop emotional resilience, coping strategies, and a sense of solidarity in facing climate change challenges together. Additionally, by fostering a sense of empathy, compassion, and interconnectedness with the natural world, education can cultivate a deep emotional connection to the environment and inspire a sense of stewardship and responsibility for its protection.

In sum, integrating an EFA model with a solutions-oriented perspective in interdisciplinary climate change education efforts in middle and high schools holds great promise for inspiring students to become active agents of change in addressing climate change challenges. By engaging students in meaningful learning experiences, promoting critical thinking and problem-solving skills, and fostering collaboration and community engagement, this model empowers the next generation to take positive action toward building a more sustainable and resilient future. Educators, policymakers, and stakeholders must continue to prioritize and support initiatives that embrace this transformative approach to climate change education, ensuring that students are equipped with the knowledge, skills, and attitudes needed to create a more sustainable world for generations to come. The final vignettes of the two projects will highlight operational barriers, solution-oriented approaches, and individual or collective actions associated with the projects.

HIGH SCHOOL VIGNETTE: CLIMATE CHANGE PROJECT

The high school setting where the Climate Change Project was implemented was in its pilot year. The program was designed to facilitate collaboration with industry partners to support student career readiness skills through focused mentoring and other events. The Climate Change Project was designed to capitalize on the structure and flexibility of local partnerships. The culminating task was for students to take action through advocacy by sharing their final projects with industry partners as stakeholders. However, this final step of the project was disrupted by the onset of early COVID policies and school closures. The schools at large encountered many operational barriers. For example, partnerships became more challenging to maintain and the project's actionable aspects fell short. Teachers were overwhelmed with navigating remote teaching, so most districts asked teachers to prioritize adapting curricular requirements to meet students' needs rather than pursue new initiatives and projects.

The solution-focused approach to the project sustained a hopeful frame of reference, which was a necessary orientation for students. There is no evidence of collective action or advocacy as a direct result of this project, but there is potential for the learning experience, knowledge, and skills to stay with the students for a long time. The PBL experience emphasized scientific literacy, interdisciplinary understanding, critical thinking, problem-solving, and proposals connected to climate action. One can hope the students who participated developed a strong disposition toward sustainable practices, environmental stewardship, and climate change advocacy through their empowering learning experience.

MIDDLE SCHOOL VIGNETTE: HUMAN IMPACT PROJECT

Students in the middle school setting were provided prompts to engage with individual action and reflection on their own sustainability practices. In various stages of the project, students were asked to complete a carbon footprint survey and water conservation survey. Additionally, students engaged in a "trash stash challenge" where they carried a bag for twenty-four hours to collect any trash items that would go to the landfill, with the goal of reducing the amount of waste each individual produced. Students took pictures for homework and weighed their bags to compete for the least amount of waste. The teacher participated and reflected with the students about how this project had prompted her to change the habit of using paper towels instead of reusable cloths or towels.

Students experienced operational barriers to sustainable action when they conducted an experiment in the cafeteria. They hypothesized that by providing directions to peers for sorting lunch waste into three bins (landfill, recycling, and compost), their proposed solution would work immediately. This led to a meaningful discussion on policy, education, and difficulty in implementation. Students concluded that additional training, visuals, and policies would help make this proposed solution plausible in their school. The failed cafeteria experiment fostered their motivation to incorporate an educational component into their policy proposal.

Students were also motivated because they would be presenting to stakeholders such as building administrators, supervisors, and watershed partners. The presentations' context was meant to help students understand how the right idea in the right hands can turn ideas into action. Students were encouraged to practice presenting at home with an adult. Sharing their learning with family was also a form of action as advocacy through science communication to engage the broader community. These opportunities empower learners and build broader engagement with sustainable practices, as parents are the decision-makers in their households.

When students presented to stakeholders, one example stood out as an opportunity for action. A student presented a multifaceted solution that included conserving electricity by reducing the length of time hallway light sensors remained on. The administrator present praised the student and their idea and shared how easy it would be for them to follow up on the suggestion. This brought a sense of joy and fulfillment to the student and the classroom community. Empowering students to communicate their ideas to local partners, boards of education, administrators, government agencies, and nonprofit organizations is a way for teachers to show they value their students' ideas. Teachers' belief in students allows them to believe in themselves. This belief strengthens motivation and sustains efforts toward climate change, sustainability, and environmental stewardship.

A Call to Action

Climate change is one of the most pressing challenges facing our planet today, with far-reaching implications for ecosystems, communities, and future generations. Addressing this complex issue requires transforming how we educate individuals about climate change and empower them to act sustainably. This chapter advocates for an interdisciplinary approach to climate change education that integrates PBL methodologies and emphasizes the importance of education for climate action. Using the imagery of Cantell et al.'s bicycle model for holistic climate change education,[46] educators and learners alike can more effectively align with these pedagogical methods for comprehensive climate change education.

Climate change is a multifaceted issue that spans different disciplines, including the sciences, social studies, ELA, mathematics, and the visual and performing arts. An interdisciplinary approach to climate change education encourages collaboration and knowledge across various fields. By breaking down traditional silos and fostering interdisciplinary dialogue, students can develop a holistic perspective on climate change and explore innovative, multidisciplinary solutions to address the global climate crisis. A comprehensive and collaborative effort across various disciplines will enhance knowledge and skill development toward climate action and sustainability.

All educators must make a collective effort to address climate change in the classroom. Educators must facilitate learning experiences that deepen understanding of climate change, as well as when, why, and how models for climate action can be created. These responsibilities fall on teachers' shoulders because they can and must create change in their classrooms, regardless of discipline. This call to action also includes fostering collaboration with partners within and beyond the classroom; from administrators and board members to nonprofits and other organizations, all community stakeholders must play an active role to be successful and effective in climate change education.

Notes

1. Emily R. Hestness et al., "Science Teacher Professional Development in Climate Change Education Informed by the Next Generation Science Standards," *Journal of*

Geoscience Education 62, no. 3 (2014): 319–29; Daniel P. Shepardson and Andrew S. Hirsch, "Teaching Climate Change: What Educators Should Know and Can Do," *American Educator* 20 (2020): 4–13.

2. National Research Council, *Climate Change Education in Formal Settings, K-14: A Workshop Summary* (Washington, DC: National Academies Press, 2012).

3. Hannele Cantell et al., "Bicycle Model on Climate Change Education: Presenting and Evaluating a Model," *Environmental Education Research* 25, no. 5 (2019): 717–31; Sarah B. Wise, "Climate Change in the Classroom: Patterns, Motivations, and Barriers to Instruction among Colorado Science Teachers," *Journal of Geoscience Education* 58, no. 5 (2010): 297–309.

4. New Jersey Department of Education, "NJ Climate Change Education Resources," https://www.nj.gov/education/climate/index.shtml

5. Asli Sezen-Barrie, Joseph A. Henderson, and Andrea Drewes, "Spatial and Temporal Dynamics in Climate Change Education Discourse: An Ecolinguistic Perspective," In *Critical Thinking in Biology and Environmental Education: Facing Challenges in a Post-Truth World*, ed. Blanca Puig and Maria Pilar Jimenez-Aleixandre (New York, NY: Springer International Publishing, 2022), 189–209.

6. Andrea Drewes, Joseph Henderson, and Chrystalla Mouza, "Professional Development Design Considerations in Climate Change Education: Teacher Enactment and Student Learning," *International Journal of Science Education* 40, no. 1 (2018): 67–89.

7. Lauren B. Allen and Kevin Crowley, "Moving Beyond Scientific Knowledge: Leveraging Participation, Relevance, and Interconnectedness for Climate Education," *International Journal of Global Warming* 12, no. 3–4 (2017): 299–312.

8. Cantell et al., "Bicycle Model."

9. Dan M. Kahan et al., "The Polarizing Impact of Science Literacy and Numeracy on Perceived Climate Change Risks," *Nature Climate Change* 2, no. 10 (2012): 732–35.

10. Cantell et al., "Bicycle Model."

11. Daniel P. Shepardson et al., "Conceptualizing Climate Change in the Context of a Climate System: Implications for Climate and Environmental Education," *Environmental Education Research* 18, no. 3 (2012): 323-352; Wayne Breslyn Et al., "Development of an Empirically-based Conditional Learning Progression for Climate Change," *Science Education International* 28, no. 3 (2017): 214–23.

12. Sezen-Barrie et al., "Spatial and Temporal Dynamics."

13. Asli Sezen-Barrie, Nicole Shea, and Jenna Hope Borman, "Probing into the Sources of Ignorance: Science Teachers' Practices of Constructing Arguments or Rebuttals to Denialism of Climate Change," *Environmental Education Research* 25, no. 6 (2019): 846–66.

14. Andrea Basche et al., "Engaging Middle School Students through Locally Focused Environmental Science Project-Based Learning," *Natural Sciences Education* 45, no. 1 (2016): 1–10.

15. Martha C. Monroe et al., "Identifying Effective Climate Change Education Strategies: A Systematic Review of the Research," *Environmental Education Research* 25, no. 6 (2019): 791–812.

16. NGSS Lead States, *Next Generation Science Standards: For States, By States* (Washington, DC: The National Academies Press, 2013), https://nap.nationalacademies.org/catalog/18290/next-generation-science-standards-for-states-by-states

17. United Nations, *Transforming Our World: The 2030 Agenda for Sustainable Development* (New York, NY: United Nations Department of Economic and Social Affairs, 2015): 41.

18. Cantell et al., "Bicycle Model."

19. Ibid.

20. Ibid.

21. Kathryn T. Stevenson et al., "Overcoming Skepticism with Education: Interacting Influences of Worldview and Climate Change Knowledge on Perceived Climate Change Risk Among Adolescents," *Climatic Change* 126 (2014): 293–304.

22. Zaretta Hammond, *Culturally Responsive Teaching and the Brain: Promoting Authentic Engagement and Rigor among Culturally and Linguistically Diverse Students* (Thousand Oaks, CA: Corwin Press, 2014).

23. Carol A. Tomlinson, *How to Differentiate Instruction in Mixed-Ability Classrooms* (Arlington, VA: ASCD, 2001).

24. Anne Meyer, David H. Rose, and David Gordon, *Universal Design for Learning: Theory and Practice* (Lynnfield, MA: CAST Publishing, 2014).

25. Meyer et al., "Universal Design for Learning."

26. Loui Lord Nelson and James Basham, *A Blueprint for UDL: Considering the Design of Implementation* (Lawrence, KS: UDL-IRN, 2014).

27. Hammond, "Culturally Responsive Teaching."

28. Laura Kieran and Christine Anderson, "Connecting Universal Design for Learning with Culturally Responsive Teaching," *Education and Urban Society* 51, no. 9 (2019): 1202–16.

29. John D. Bransford, Ann L. Brown, and Rodney R. Cocking, *How People Learn*, vol. 11 (Washington, DC: National Academy Press, 2000).

30. Cindy E. Hmelo-Silver, "Problem-Based Learning: What and How do Students Learn?" *Educational Psychology Review* 16 (2004): 235–66; Linda Torp and Sara Sage, *Problems as Possibilities: Problem-Based Learning for K-12 Education* (Arlington, VA: ASCD, 1998).

31. Basche et al., "Engaging Middle School Students."

32. Joseph S. Krajcik and Charlene M. Czerniak, *Teaching Science in Elementary and Middle School: A Project-Based Learning Approach* (New York, NY: Routledge, 2018).

33. Violet H. Harada et al., "Project-Based Learning: Rigor and Relevance in High Schools," *School Library Management* 157 (2015): 14–16.

34. Basche et al., "Engaging Middle School Students."

35. United Nations, "Transforming Our World."

36. Kieran and Anderson, "Connecting Universal Design."

37. Ibid.

38. Anja Kollmuss and Julian Agyeman, "Mind the Gap: Why do People Act Environmentally and What are the Barriers to Pro-Environmental Behavior?" *Environmental Education Research* 8, no. 3 (2002): 239–60.

39. Cantell et al., "Bicycle Model."

40. Doug Lombardi and Gale M. Sinatra, "Emotions about Teaching about Human-Induced Climate Change," *International Journal of Science Education* 35, no. 1 (2013): 167–91; Maria Ojala, "Hope and Climate Change: The Importance of Hope for Environmental Engagement among Young People," *Environmental Education Research* 18, no. 5 (2012): 625–42.

41. Sezen-Barrie et al., "Spatial and Temporal Dynamics."

42. Monroe et al., "Identifying Effective Strategies."

43. Allen and Crowley, "Moving Beyond Scientific Knowledge"; David Hicks, *Educating for Hope in Troubled Times: Climate Change and the Transition to a Post-Carbon Future* (London: Institute of Education Press, 2014); Ojala, "Hope and Climate Change."

44. K. C. Busch and Jonathan Osborne, "Effective Strategies for Talking about Climate Change in the Classroom," *School Science Review* 96, no. 354 (2014): 25–32.

45. Cantell et al., "Bicycle Model."

46. Ibid.

CHAPTER 7

Climate Change Education in the STEM Classroom

Rachel DiVanno, Kelly Stone, and Melissa Zrada

This chapter describes how to use the interdisciplinary approach of science, technology, engineering, and mathematics (STEM) teaching to embed climate change into problem-solving lessons, specifically in the middle school classroom.

One strategy for teaching and learning about climate change in the STEM classroom is by taking an interdisciplinary approach. Rather than teaching each STEM subject separately in a siloed way, topics are integrated to create a cohesive learning experience, one that mirrors the ways professionals work in the real world, thereby creating a more authentic learning environment for students at all grade levels. The International Technology and Engineering Education Association defines *integrative STEM education* in the following way:

> the application of technological/engineering design based pedagogical approaches to intentionally teach content and practices of science and mathematics education through the content and practices of technology/engineering education. Integrative STEM Education is equally applicable at the natural intersections of learning within the continuum of content areas, educational environments, and academic levels.[1]

The New Jersey standards to support climate change learning are divided into grade bands K–2, 3–5, 6–8, 9–12, as well as the following subject areas:

- Visual and Performing Arts;
- Comprehensive Health and Physical Education;
- Science;
- Social Studies;
- World Languages;

- Computer Science & Design Thinking; and
- Career Readiness, Life Literacies and Key Skills.[2]

For each standard, "core ideas" (or "enduring understandings" for Visual and Performing Arts) and performance expectations are provided. The science standards are grounded in the Next Generation Science Standards[3] (NGSS), but unlike the NGSS, they include ideas explicitly related to climate change, beginning in kindergarten rather than delaying until middle school. This ensures that middle schoolers exploring climate change are not doing so for the first time and will come to our classrooms with some background knowledge on the topic.

One key component of the NGSS is the inclusion of engineering content within science and explicit connections to mathematics and English-language arts through shared language in the practices for each discipline. Appendix D of the NGSS offers a Venn diagram delineating how science content is integrated with these other important content domains.[4] This purposeful integration allows science classes to serve as ideal spaces to teach using a STEM-based interdisciplinary approach. Similarly, New Jersey's Computer Science and Design Thinking standards offer opportunities to apply the engineering design process across disciplines. By using standards in both areas, teachers have a clear framework for interdisciplinary teaching and learning.

Core ideas in middle school science standards (grades 6–8) span topics that include ecosystems, earth cycles, weather and climate, and human activities. Core ideas in computer science and design thinking at this level cover topics such as the design process and technological systems. Career readiness, life literacies, and key skills also include topics relevant to STEM, including information and ethics.

The New Jersey standards to support climate change learning include the seven NGSS crosscutting concepts in connection with core ideas across disciplines. These crosscutting concepts include (1) patterns; (2) cause and effect; (3) scale, proportion, and quantity; (4) systems and system models; (5) energy and matter; (6) structure and function; and (7) stability and change. For example, students may discuss patterns when making observations about weather data, cause and effect when learning about human activities and their impact on the environment, and systems and system models when brainstorming solutions to real-world challenges. While these examples are by no means exhaustive, they exemplify how the New Jersey standards to support climate change learning support the crosscutting concepts.[5]

Problem-based learning is one interdisciplinary approach often used to integrate STEM into the classroom. Chapter 13 of this book provides specific details on how to integrate this approach into the curriculum. However, there are other ways teachers can use an integrative STEM approach in their work by presenting climate change content in middle-school classrooms. Broad ideas and suggestions for this kind of engagement are described in the next section of this chapter, followed by one example of using standards to guide multifaceted middle school STEM instruction around climate change and solutions.

Climate Change as a Way to Engage

Student engagement can be tied to academic achievement; when students exert effort in their learning, it often results in higher grades. This effort goes hand-in-hand with the "buy-in" or motivation for students to care about and take ownership of what is being taught. Research shows that student engagement allows students to delve into a topic, draw connections about prior knowledge, and uncover pathways between ideas.[6] Climate change is a logical topic for educators to use to promote engagement in the classroom, because it activates the emotional side of students' care for the planet, places of importance, animals, and people.[7] Since everything students have experienced in their lives is technically part of the environment, educators can use what students care about as examples when they teach about climate change.

Further, climate change is an interdisciplinary topic that allows students to use real data and situations to create claim-based statements and solutions. Claim-based statements can be used in subject areas from ELA and social studies to math and science. Learning how to interpret data and draw conclusions from a dataset is a necessary skill students learn in every subject area. Climate change can be used outside the science classroom as a bridge between different subject areas to help students practice making evidence-based claim statements and analyzing data. It is a broad enough topic that educators can tailor the ideas to place-based learning opportunities. Students feel empowered, interested, and in control of their learning when examples are brought to them about a location in which they live.

Students can also use critical thinking skills to analyze information and make informed decisions based on the data presented to them. Students are often tasked with being critical thinkers when educators prepare them to face problems, jobs, and situations that have never existed before. To prepare students for these kinds of challenges, we must encourage students to think critically and make inferences about the future that will eventually inform their decisions. Teaching climate change is a practical method to build middle schoolers' critical thinking skills because students are currently experiencing the effects of climate change and can draw on their own lived experiences in the process of learning about science and making predictions. This critical thinking about causes and effects leads students who began as learners to end as problem solvers.

Critical thinking about climate change and its solutions can help students establish a hopeful mindset about the future. If a doomsday approach is used, students may shut down and feel helpless. To engage students on the topic of climate change, it is necessary that at the end of the lesson, they have a sense of action or control. For example, teachers can ask students to calculate individual carbon footprints and reflect on how changing certain actions in their own lives makes a difference to their footprints. This activity allows students to feel in control and gives them actionable steps that could lead to real change. Students can also talk about their communities and areas in which they could help improve the environment. Sometimes when teachers introduce government policies or big corporations to students, climate change can seem too large, too adult, or too complex to tackle. Ending each lesson with a hopeful approach can help inspire students to act.

Climate change activism could also encourage students to take agency and advocate for solutions. Before we introduce environmental activism to them, students may only associate activism with certain events they have learned about in social studies class. Allowing students the opportunity to develop their relationship with activism at an age where they are not in total control of their choices around diet or spending can lead to environmental stewardship.[8] Students can choose to have meatless Mondays, organize a walk-to-school day, or collect recycling in the school cafeteria to take action to mitigate climate change. Giving students examples of real young people who have made names for themselves through their activism can lead to students becoming leaders in the climate change movement. Another strategy that can help is introducing students to the United Nations Sustainable Development Goals (SDGs) because it demonstrates that there is a global collective commitment to working together for a better future. This helps support a hopeful outlook while giving students resources to explore the biggest issues the world faces and actions that are already being taken.

The emotional connection involved in caring about where you live helps engage students in the conversation about climate change and its impacts on the present day and the future. Using a place-based learning approach, where teachers pull examples from where students live, provokes more empathy for the topic.[9] Several resources provide data from places all over the world to give students a better understanding of what is happening in their community. Consider having students explore the National Oceanic and Atmospheric Administration (NOAA), Environmental Protection Agency (EPA), and NASA websites. They can observe brownfields in their state through the EPA, study weather and climate maps on the NOAA site, and analyze atmospheric indicators of climate change on NASA. All of these sites also provide sections for educators where teachers can access data and resources for free to use in their classrooms.

Science educators often use phenomena-based instruction to engage their students. The teacher presents a phenomenon, and students develop questions and identify areas for further exploration. This method puts students in control of their learning and helps them move from learner to active participant in sense-making on a particular topic. In this approach, students identify and uncover the "big ideas" relevant to a given example, instead of teachers presenting all the information, as in traditional instruction. Phenomena-driven instruction has received increased attention since the adoption of the NGSS because they can provide contexts for learning and require students to problem-solve rather than memorize. Broadly, climate change offers many opportunities to present phenomena in the classroom. Teachers can share phenomena related to the effects of climate change such as rising sea levels, ocean acidification, and the replacement of native plants by invasive species; then, students can grapple with ways to analyze and interpret these events. Eventually, students are able to develop conclusions about why these events are occurring without the teacher lecturing about the topic.

Climate change is an overarching thread that can be explored using various kinds of thinking across academic disciplines and domains. Student interests can be piqued through climate change investigations, which can lead to further engagement. Whether students are learning about how instruments have been used to alert or tell stories about the weather or how art is used to send emotional messages about climate-change activism, students can analyze climate graphs and data in math class or practice evidence-based writing in ELA. The important takeaway is that all teachers feel comfortable using climate change in their classrooms as a method to instruct because it gives students multiple perspectives and viewpoints through which to engage in the topic.

Using NGSS Standards to Bring Climate Change Lessons into Middle School Classrooms

Though the opportunities for using STEM to engage middle schoolers in learning about climate change are plentiful, it can be challenging to identify starting points. Here, we offer suggestions for starting with the NGSS standards to structure meaningful learning about climate change in the middle grade.

One consequence of climate change that all humans experience is the devastating effects of natural hazards. The following standard provides clear guidance on how to guide students in unpacking these events: "MS-ESS3-2: Analyze and interpret data on natural hazards to forecast future catastrophic events and inform the development of technologies to mitigate their effects."

Students can access regional weather data and begin by categorizing and graphing historical catastrophic events and natural disasters prevalent in their region. After defining categories and observing timelines, students can analyze these data and make suggestions to inform future decision-making. For example, students can look at the data and make suggestions about how home and business structures could be impacted and work to create a solution to safety concerns in natural hazard and disaster situations.

Human population and overconsumption of natural resources over the past century have exacerbated the effects of climate change. The following standard provides a framework for exploring this cause of climate change for students: "MS-ESS3-4: Construct an argument supported by evidence for how increases in human population and per-capita consumption of natural resources impact Earth's systems." Using data visualizations, such as the tools suggested in Chapter 9 of this book, can help middle-schoolers conceptualize their understanding of the relationships between human population, fossil fuel use and deforestation, and the impacts of resource overuse on atmospheric temperature. Visualizations can make the mathematical manipulations needed to understand these relationships concrete.

Other related standards such as the ones below give students a chance to use an inquiry-based problem-solving approach to explore causes of climate change: "MS-ESS3-5: Ask questions to clarify evidence of the factors that have caused climate change over the past century." When the standards themselves require students to craft these kinds of questions, students can take ownership of their own learning experiences. One project-based learning investigation could involve having students compare human populations in local regions to climate change impacts in those areas and construct a solution to reverse impacts where population increases are more prevalent.

When examining differences in climate change impacts across the globe, biodiversity loss will inevitably come up as a key consequence. The following standard requires teachers to consider STEM-based design solutions to help mitigate this problem: "MS-LS2-5: Evaluate competing design solutions for maintaining biodiversity and ecosystem services." Biodiversity has changed in many parts of the world, especially where overuse of resources and the resultant climate change has caused treacherous impacts to land and water. Students can use data visualizations to define areas where this has had a direct impact on the availability of food sources, shelter, and clean water for both humans and animals. Focusing on water in particular can provide opportunities for connecting middle schoolers' STEM learning experiences to the United Nations' SDGs, Goal #6 in particular: "Ensure availability and sustainable management of water and sanitation for all." Focusing students' attention on this important global goal adds purpose and relevance to STEM learning. Students can collaborate using tangible materials and technological tools to design a solution to accessible clean water and explain how human consumption and access to water affect biodiversity. This focus also provides opportunities for further interdisciplinary collaboration.

The following social studies standards open the door for teachers to approach learning from an interdisciplinary perspective:

- "7.1.NM.PRSNT.6: Name and label tangible cultural products associated with climate change in the target language regions of the world." The impacts of climate change range in severity for different regions of the world, and as a result, they can and will be communicated differently. Emphasizing the importance of communicating across different formats can help students develop agency and advocate for widespread climate change solutions. "7.1.NM.IPRET.5: Demonstrate comprehension of brief oral and written messages found in short culturally authentic materials on global issues, including climate change." Communication entails more than just written and verbal messages. Additionally, the next standard provides opportunities for students to consider artworks and media messaging. "1.2.8.Re7b: Compare, contrast and analyze how various forms, methods and styles in media artworks affect and manage audience experience and create intention when addressing global issues including climate change." Middle school students often engage with the world through visual and gamified versions of society that connect real-world events, ideas, and possible projections into movies, social media content, and games. When teachers encourage all forms of messaging, it allows students to find solutions that resonate with them and their preferred modes of communication.

Middle school STEM is rich with opportunities for connecting students to hopeful climate solutions. Valuing interdisciplinarity, problem-based learning, collaboration, and multiple media formats can help empower students to take agency.

Notes

1. International Technology and Engineering Educators Association, "Integrative STEM Education," https://www.iteea.org/integrative-stem-education.
2. New Jersey Department of Education, "New Jersey Climate Change Education: Grade Band Learning Goals," https://www.nj.gov/education/climate/learning/gradeband/
3. National Research Council, "High School - Physical Sciences," in *Next Generation Science Standards: For States, By States* (Washington, DC: The National Academies Press, 2013), https://www.nextgenscience.org/dci-arrangement/hs-ps.

4. National Research Council, "Appendix D: All Standards All Students," *Next Generation Science Standards: For States, By States* (Washington, DC: The National Academies Press, 2013), https://nextgenscience.org/sites/default/files/Appendix%20D%20Diversity%20and%20Equity%206-14-13.pdf.

5. National Research Council, "Appendix G: Crosscutting Concepts," in *Next Generation Science Standards: For States, By States* (Washington, DC: The National Academies Press, 2013), https://www.nextgenscience.org/sites/default/files/Appendix%20G%20-%20Crosscutting%20Concepts%20FINAL%20edited%204.10.13.pdf.

6. Samantha Frost et al., "Student Engagement and Environmental Awareness," *Environmental Humanities* 14, no. 1 (2022): 219-233, https://read.dukeupress.edu/environmental-humanities/article/14/1/219/294322/Student-Engagement-and-Environmental-AwarenessGen.

7. Ibid.

8. Action for the Climate Emergency, https://acespace.org/.

9. Kelly Kleinertz, "Using Place-Based Learning to Teach Middle School Students About Climate Change," *Edutopia*, June 5, 2023, https://www.edutopia.org/article/place-based-learning-climate-change/.

CHAPTER 8

Climate Change Connections
THE POWER OF KNOWING AND DOING, TOGETHER!

Cari Gallagher and April Oliver

As the issue of climate change continues to permeate our world, it has become vitally important to create and strengthen connections through the public school curriculum to support environmental education and civic responsibility. Lately, climate change has become something of a controversial or divisive topic, and it appears some educational institutions are reluctant to address it or may favor using a "hands-off" approach to teaching it. Recent legislation in New Jersey requires that K–12 public schools include the 2020 standards to support climate-change learning in lessons across the core content areas. To that end, questions arise about the quality of those connections and about the efficacy of addressing climate change both in the classroom and at home. How practical is it to include these new standards in content areas besides science? Are students included in open, constructive climate change discussions in their classrooms? On the home front, is there a dialog in which parents include their children in conversations about our changing earth? Do parents know how to engage their children in age-appropriate climate change conversations? Climate change can be addressed in families with young children in the simplest ways, such as using everyday weather examples. Open discussions where children and adults can share ideas, opinions, and experiences can also create pathways for a sustainable future.

Many middle and high school students are passionate about environmental justice; youth campaigns for action are growing both in number and popularity. However, fewer supports are in place for elementary school-aged children to become engaged with this movement. Younger students need guidance from educators and community members to foster their interest in the natural world and to examine the broader detrimental implications of climate change in an age-appropriate way. With the introduction of New Jersey's *Career Readiness, Life Literacies and Key Skills* standards for kindergarten through second grade, younger children are now able to investigate and participate in research about climate change, learning various ways to take civic action. For example, standard 9.4.2 describes a way "Young people can have a positive impact on the natural world in the fight against climate change."[1] Providing an educational framework in schools that can be unpacked at home with families can lead to positive changes in our environment and community. This framework needs to begin at the state level, which can trickle down to individual cities and towns to individualize climate change education relative to their school communities.

New Standards: Now What?

The New Jersey Department of Education (NJDOE) has designed interdisciplinary New Jersey student learning standards about climate change to be used in public K–12 schools. Sample climate change units are also available on the NJDOE website in a number of grade bands. The science learning standards, grounded in the Next Generation Science Standards, guide teachers using foundations of science and engineering practices, disciplinary core ideas, and crosscutting concepts that relate specifically to performance expectations and connections and provide a good starting point for teaching about climate change. However, educators must now synchronize learning objectives for climate change instruction using grade band-specific ideas across six additional content areas: visual and performing arts, comprehensive health and physical education, social studies, world languages, computer science and design thinking, career readiness, life literacies, and key skills. The NJDOE has also codified a climate change companion guide called STAMP (Standards Transparency and Mastery Platform), which can be applied in English, language arts, and mathematics within the K–8 New Jersey student learning standards. The goal of the STAMP guide is to support teachers in incorporating climate change education across multiple content areas to foster future generations of students.[1] The STAMP guides were released in 2023 and will be implemented in the 2024–25 academic year. These initiatives have empowered teachers to begin transferring climate change concepts into math by using real-world examples and in language arts with the support of various book genres and current events.

Teachers have also begun to revise traditional content-specific lesson plans with the integration of climate change concepts, vocabulary, and visual representations. Many specialist teachers, such as those teaching art, music, and physical education, have expressed concerns about having "something else to teach" along with their required curriculum, so ensuring resources, professional development, and a clear understanding of teaching climate change using developmentally appropriate practices is essential to their students' success. Reassuring all teachers that these standards are intended to be implemented in multiple content areas is essential to set the tone for inclusive climate change lessons that address the climate crisis.[2]

Though teachers have additional standards to use as a framework, they are also tasked with producing effective, developmentally appropriate lessons, often without formal professional development or guidance. One key to ensuring teachers have access to professional learning is making them aware of a multitude of free, reputable online resources. Teachers can attend professional development on climate change concepts hosted by a number of trustworthy organizations, including NASA, SubjectToClimate, The Climate Change Education Exchange, Rutgers Climate Institute, and through the National Oceanic and Atmospheric Administration.

Top-Down Leadership

Like any other educational initiative, having the support of district administrators is the key to successfully implementing climate-change education. Many districts have already started showing support by appointing Green Team leaders, as is the case in our district of Lawrence Township in the central region of New Jersey. However, district leadership must prioritize funding, programming, and essential professional development to deliver climate change instructional practices effectively. School districts may presently have programs in place in language arts, math, and social studies that teachers can use as a starting point for K–4 climate change lessons. Simply taking the time to look for places where climate change fits with the existing curriculum can benefit teachers across grade levels. Other educational institutions may need additional guidance, and administrators can write federal grants and collaborate with their local educational associations for financial support. Grants and local organizations can provide supplemental resources for students, such as multiple copies of books for literature clubs, posters for display, gardening expansion, and climate change presenters for school assemblies to show students the importance of sustainable practices to empower our earth.

One central location all students utilize is the school cafeteria. This is a great starting place for administrators to recalibrate their schools' culture to become more sustainable. Promoting recycling, composting, waste reduction, and civic responsibility during lunchtime is an effective way to develop healthy habits and support climate-change solutions. Financially, there may be issues with hiring additional personnel to help oversee these initiatives, so asking for school volunteers or assigning school ambassadors can strengthen these best practices.

School to Home

As climate change themes continue to be addressed in K–12 curricula, young students in kindergarten may start exploring the basics of weather and long-term climate patterns. Much of this information can powerfully impact younger children. Even though many families include their children in climate change discussions at home, it is unclear to what extent they participate in larger environmental conversations. A recent study found that parents across New Jersey requested further support from teachers and schools to help guide home-based conversations about climate change.[3] Parents and guardians need to remember that children can grasp the impact of climate change as long as it is portrayed in a developmentally appropriate manner. Students will make connections if they understand the basics of everyday weather on climate. Whatever families' political views, educators can point to factual longitudinal scientific data that explains how and why climate change is happening.[4] Providing families with basic knowledge is the key to acceptance, which may lead to productive partnerships between families and educators.

A starting point for transferring climate change instruction from schools to homelife is to create a fact-based foundational guide of this worldwide problem. Many guides are also available online, including one published by NASA.[5] Diplomatically providing parents and guardians with evidentiary resources and sharing factual explanations of interdisciplinary climate change lessons can be very helpful. Even though families carry different political viewpoints, scientific data is concrete and should be available for the community to support students' learning.

Social-Emotional Learning

There are direct connections between climate change instruction and students' social-emotional growth in K–12 schools. Many districts use character education points of learning, where students are provided with scenarios, stories, and instruction that highlight decision-making, advocating, and being responsible. Educating students about the impact of human behavior is a core component of understanding the climate change crisis. These climate-change conversations can ignite empathy and compassion at many levels in young students, so being mindful of the delivery of environmental events is essential for students' emotional stability. Curriculum revisions typically take place in districts every three to five years, and climate change is now woven into curriculum guides. As students become more knowledgeable about the facts of climate change, they are likely to realize that their generation could be affected by it the most.

Throughout social-emotional learning, educators need to discuss solutions such as renewable energy options, which can support students' further action and environmental justice involvement. Helping students understand that no one wants to intentionally hurt our earth is a first step; as students grow and move on to middle and high school, they can grapple with larger issues, such as how economics plays a role in decision making to better our environment, as well. At the elementary level, the goal should be to highlight issues and actionable solutions without frightening young students or causing them "eco-anxiety."[6] Talking about solutions and outcomes and sharing creative problem-solving ideas should leave students feeling empowered and positive about their influence on the future.

Forward-Thinking

The concept of teaching students to think about the "big picture" is sometimes difficult for young learners to grasp. It is not always natural for students to think past a problem without emotions getting in the way. Children are often focused on their immediate present, so using something as simple as the calendar or the daily weather outside could be a starting point for climate change connections. During the COVID-19 pandemic, field trips for schools were temporarily placed on hold. Students had been going to museums, art exhibits, and theaters for field trips, which exposed them to opportunities to connect with nature, participate in the environment, and interact with institutions supporting climate change education. Since the return from COVID-19 restrictions, schools have made more of a concerted effort to promote field trips and other engaging activities such as community events and service-learning.

School administrators began to realize climate-change education thrives with enriching programs and field trips where these concepts can be extended. Some students may have difficulty using forward-thinking in the classroom, so field trips are a unique way for them to practice it in a real-world setting. These educational outings begin with buses, which open the door for conversations about emissions and electric conservation. Students can learn how human activity can be part of the problem and the solution for climate change. Visiting museums and conservation centers and exposing students to activism that supports climate change in a positive way can empower students to be climate stewards. Students can make personal connections from decisions that impact climate change. For example, aviation is responsible for two to three percent of global CO_2 emissions and air travel is expected to double in the next decade.[7] Explaining the long-term impact on fuel emissions from traveling can encourage students to become involved in future decision-making with their family, friends, and communities.

A Patient Mindset

Patience may not be the first thing you think of when visualizing young children in a classroom setting, and for good reason. Learning to wait, share, and listen are not always innate traits, and they need to be built from repeated daily practices. At our school, Lawrenceville Elementary School, we use the model of *a growth mind-set versus a fixed mind-set*.[8] It is a great start for students to learn what they can achieve at any age.

For instance, one of the most common reasons people do not recycle properly is because of all the extra steps it involves. Recycling a peanut butter container made of plastic is a good start; however, there cannot be any peanut butter left inside it. Are people thoroughly washing out their containers before recycling? Placing a plastic bag of water bottles in a recycling can is a responsible action, but the plastic bag cannot go into the water-bottle-processing area, or it will gum up the machination. All of these extra efforts require patience and perseverance, which is something even adults struggle with. Patience also continues to be an important attribute needed for long-term follow-through regarding the climate change crisis. Efforts made today may not pay off for decades, which is a complicated concept for young minds living in a world filled with instant gratification.

When we talk about helping our environment, we need to be mindful of the audience, which may vary across suburban, rural, and urban school communities. Many families may be fortunate to have financial support that allows them to focus on proactive behaviors like water regulation, solar panels, proper recycling, and composting. Other families are not as fortunate and may not

Fixed Mindset	Growth Mindset
I CAN'T do that	I can't do that YET
Earlier failures or anticipated failure indicate that we can't do it.	Earlier failures or lack of familiarity indicate that we can't do it yet.
'I know I can't do that'. 'I've tried it before and proved I'm hopeless at it'.	'I know I can't do it yet but I'm willing to have a go'. 'I'm hopeful I can do it better next time'.
Saying 'I can't do it' as an excuse for not joining in.	Saying 'I can't do it yet' to reflect the possibilities of future success, and to signal a willingness to try.

Figure 8.1 Growth Versus Fixed Mindset

make climate change conversations a priority at the dinner table. Due to diverse socio-demographic and cultural factors, personal beliefs and attitudes about the importance of climate change may vary.[9]

Collaborative Teaching as a Resource

One easily implemented and cost-effective way to support classroom teachers in reaching cross-curricular standards is to encourage teacher collaboration. The school librarian (or media specialist) is an excellent resource to co-teach lessons with, help with pairing resources and lessons, and direct enrichment projects that address climate change. Research-based activities can include projects such as reading and writing about a variety of animals in different habitats. Students can then research how climate change could impact the biomes where those creatures live. A second-grade unit taught collaboratively in the library media center might include reading about animal habitats and describing the ways climate change, pollution, and deforestation apply pressure on animals to migrate to new locations, seeking new food and water sources. Students could then be shown a series of photographs of "animal crossing" walkways and overpasses from around the world that have been constructed to allow animals safe passage to new areas of land.

Many school librarians have also developed collections of nonfiction books that highlight recycling, pollution, global warming, animal conservation, and biographies of nature conservationists. These nonfiction books can be paired with language arts lessons the students have already completed with their teachers. By contrast, fictional material that touches on these same themes can be used to make literary connections to what students have learned in their science units. Teachers and students could also use online databases that have been vetted by the school librarian in their own classrooms for research purposes. If the school librarian is hindered by time constraints, a quick read-aloud book on any of these topics during the library period would help enhance those connections for all students.

Makerspaces

Makerspaces allow students to undertake purposeful, thoughtful activities to help reinforce what they've learned about in their classes. In many ways, it's the "do" part that comes after *seeing* and *knowing*. As much as elementary school students enjoy an open-ended non-structured period for free-choice activities, we have observed in our school that the most successful makerspace periods occur when students are given parameters and objectives with which to work;

they create meaningful content from that. For instance, a second-grade lesson referenced earlier about climate change impacting animal migration could culminate with students building their own animal crossings using Lego Bricks, straw connectors, magnetic blocks, and other materials.

Design thinking in education[10] is a process by which students are presented with a problem they must collaboratively discuss to solve. They will often use blueprint paper to plan out their thinking in advance and can draw their ideas and make lists of the material they want to use. Design thinking is a perfect process to use during makerspace time, whether it's in the library or classroom. Teachers and/or librarians can give groups of students scenarios rooted in climate change, such as the following: "The summers in New Jersey are getting warmer. How does that affect the farm animals living outside? Design a new chicken coop that allows for lots of air circulation, so the chickens on the farm can keep cool."

After the initial planning and discussion is complete, students are encouraged to work within a group to construct projects using materials of their choice. The activity ends with the group explaining what their solution model was and how they constructed it. Rubrics help teachers assess how successful the activity was and gives students a chance to reflect on their role in the design thinking process. Table 8.1 provides an example of a rubric used for a design thinking project completed at our school.

Makerspaces can be as simple, small, mobile, or elaborate as teachers want them to be. Many schools have a makerspace included in their libraries as a gathering point for donated or reclaimed material, prefabricated kits, iPads, and other technology. However, maker-carts offer an inexpensive, low-tech alternative to a formal space. Maker-carts can be purchased or repurposed from old audio-visual carts and can hold all manner of material students can use for their

Table 8.1 Climate Change Chicken Coop Project Rubric (The expectation is to achieve 1-3 for 90% of the class.)

Not Engaged (0)	Engaged (1)	Engaged with Curiosity (2)	Engaged with Curiosity and Higher Order Questions and Cross-Curricular Connections (3)
Student did not identify an issue or problem.	Student pre-planned their design (with blueprint) either individually or collaboratively with group.	Student was engaged collaboratively with group and was hands-on in building/construction.	Student was immersed in the activity and included a post-analysis of their design and experiment.

Figure 8.2 A koala crossing in Australia providing safe movement over a highway inspires students to create their own animal crossing in the makerspace.

projects; they also provide the convenience of being transported from classroom to classroom. Whatever the makerspace technically is, it is important to offer younger students creative, thematically inspired activities. Engaging activities will be memorable for them and will further cement their understanding of abstract concepts, such as climate change.

Special Spaces

At the elementary-school level, visual reminders of school rules, mottos, and positive affirmations can be seen in entrances, hallways, and hanging on doors and windows. A natural extension of those decorative devices could be special small spaces where students can read, relax, or energize their minds and bodies while pondering the natural world around them. One example is to install a "climate change corner" in the classroom, hallway, or library where students can read magazines and books about climate conservation and advocacy. Working with a very limited budget, a "climate change corner" might consist of a green rug placed in a corner along with a wooden bookshelf holding reading material. If more funds are available, pillows that look like logs could be added, along with an aero-garden growing greens, giving the illusion of bringing the outside world

Figure 8.3 Students collaborate on climate change projects in the makespace.

inside. At our school library, a list of reading material is housed in the climate change corner, including age-appropriate biographies of Greta Thunberg and other environmentalists. Students can use the corner during "free reading" periods, as a special calm-down spot, or even as an incentive during a Fun Friday. There are many ways to utilize creative spaces.[11]

Another item used in a "special place" at our school is an eco-charger bicycle. It's a stationary bike students use for "read and ride" breaks. Its purposes are multifold: It allows students to burn off excess energy while reading a book propped up on the bookstand attached to the bike. At the same time, children can help power their school building by generating watts of energy. The eco-charge bikes plug into any normal wall outlet, while the bike can be set up as far as fifteen feet away. Energy then goes back into the building. Most kids can generate fifteen to thirty watts from pedaling, while they read for up to fifteen minutes![12]

Civic Responsibility & Service-Learning

Service-learning is an integral part of most elementary schools' educational programs and is often led by the guidance counselor. However, there are many opportunities to combine green and sustainable initiatives within those service-learning experiences. For schools that have a garden with vegetable harvests, there are abundant opportunities to learn in an outdoor classroom about soil, clean water conservation, food production and harvesting, and how we are all connected with the living world. For example, a traditional service-learning project at our elementary school focuses on collecting donations for a Thanksgiving food drive, but it could also incorporate themes of how we can and should ensure the future of healthy soil and water in our environment. Students can discuss how food waste is a serious problem, especially because in every community, there are people who often don't have enough to eat. Students can also learn about how access to drinkable water and nutritious food is imperative for everyone. In addition to having students eat a salad at lunch time from the school's harvested crop, an extension lesson could investigate why and how donating grown vegetables to a food bank or pantry is important. Encouraging all students to bring reusable water bottles to school and refill them at the fountain stations instead of bringing in plastic bottles could also tie into conversations about sources of fresh water. Fresh water is not an infinite resource, and can become at risk of being affected by the warming climate. Centering the student as the first line of action for civic engagement is vital and helps students feel empowered to accept personal responsibility.

Many service-learning projects also focus on volunteerism to clean the school gardens, recycle material both at home and during school lunch times, and collect recycled items for community members in need, such as gently used coats. Recycling material to better our local communities goes hand in hand with improving the planet's environment. Students can make connections between the micro-level of their local community to the macro-level of our planet as a whole. Discussions of how pollution and global warming impact not just our earth but also our local environment is important; it may start in the classroom as a climate-change standard but continue as a service-learning project.

Community Partnerships and Recognition

Community partnerships can be formal or very informal. Communities and public schools can intersect to promote fun, inclusive programs and displays from which all students benefit. Our school is currently housing a "Bee Tree" display, created by the Landscape Committee of the Lawrenceville Main Street Organization. This tree was entered in the 2023 Festival of Trees at Morven Museum in Princeton, New Jersey. The title of the tree is "The Beauty of Bees" and it tells the story of bees and their social structure, their amazing behavior, and their important pollination work. The decorations include bees, beehives, honey jars, and flowers. In addition, true-or-false fun facts about bees are displayed on honeycomb discs, making the tree a fun "quiz" with which students can engage.[13]

For more formal recognition, there are many ways public schools can get involved in sustainable practices. Currently, 67 percent of New Jersey public schools have achieved either bronze or silver certification from Sustainable Jersey for Schools. This nonprofit organization recognizes schools' and districts' efforts and proactive measures to reach sustainability goals. The New Jersey Education Association partners with Sustainable Jersey to support projects and initiatives that positively affect students, staff, and communities surrounding many New Jersey schools.

The extensive application process covers many areas of sustainable development from curricular connections to climate mitigation, community outreach, and food and nutrition. In many ways, it provides a framework for sustainability that could be a model for schools at the national level. Schools (and districts) must take a keen inventory of their carbon footprints and document their best practices from green cleaning to school culture and climate. It is an exhaustive but rewarding endeavor that can open up the schoolwide community to a conversation about the district's sustainable direction: Where are we headed? What can be improved upon? What are we doing well and where are we lacking? All these conversations can be had across the district and help focus the future goals of the schools and their communities. Reviewers of schools' applications provide detailed feedback and encourage schools to resubmit evidence that may need further elucidation. It is a collective effort, and achieving certification is an important recognition. Schools and districts must reapply every three years, which helps encourage new and enhanced sustainability efforts.

Family Engagement

Elementary school students are naturally active learners on their school campuses. Students can typically experience recess outside, nature walks, stream strolls, and garden lessons, depending on their school's location and funding. However, how can students also use social-emotional learning to take climate action at home? How can families get involved in making school-to-home-life connections about the climate?

School districts host back-to-school nights, math and science nights, and book clubs for students and parents in many elementary schools. They may also occasionally hold community conversations; however, climate change is sometimes highlighted only in the spring around Earth Day and Arbor Day. Using April as National Earth Month is helpful to solicit information at parks, nature centers, and watersheds, but those efforts appear to dissipate until the next spring comes. Touching upon climate change intermittently is not a solid model of educating youth on changes to the climate.

In schools, students can benefit from seeing the effects of climate change daily throughout all seasons, as it can serve to spark conversations at home. Climate change connections can strengthen viewpoints and encourage further research and action to help our planet. Even though school administrators have big ideas to carry on with climate change initiatives, they tend to lose priority to core content being taught and assessed in the public school system.

Teachers and school personnel can look to local and federal grant resources to help strengthen and provide their school communities with family engagement activities that support climate change education in elementary schools. While it can be very difficult to retain teachers and administrators for afterschool and evening programs in school districts, if climate change instruction is made an integral part of the school's mission statement and a priority in every classroom, students can grasp environmental concepts through active stewardship.

Summary

The Department of Education continues to support New Jersey public schools with interdisciplinary standards to support climate change learning and sample units to help educators frame lessons for their students. The NJDOE has aligned climate change education lessons by grade band with enduring understandings and performance expectations.

New Jersey teachers have an effective reference for climate-change instruction, which can be shared with parents and school stakeholders to increase climate action and awareness through instruction. K–4 students would benefit from lessons designed through a constructivist lens, using real-life examples. As students become more familiar with the causes of climate change, the impacts of human activity, and the increase of fossil fuels and emissions in our environment, they can make connections to their own experiences.

Keeping diversity at the forefront of education, teachers can share books with parents to continue embedding healthy habits for students to prioritize their well-being and decision-making and become upstanders in their communities, regardless of their accessibility to climate aid resources. Since 2013, global climate change has been included in the science curriculum worldwide, using core ideas and content-specific domains, which has provided teachers with a solid foundation of science lessons.

Schools can promote the message of "knowing and doing" by integrating climate action initiatives into all subject areas in the curriculum. Starting at the elementary level, students can develop positive habits for themselves and the planet, which can be shared with families at home. School Green Teams typically reinforce districts' sustainability efforts, inviting parents, administrators, stakeholders, teachers, and students to collaborate on practices that empower climate action in the school community.

Working with the board of education is also a productive way to get academic and financial support to build climate change libraries, develop sustainable gardening programs, and hire additional personnel to implement these school practices, which ultimately lead to family engagement. Seeking out additional funding for Title I schools is also worthwhile to finance these goals.

Notes

1. New Jersey Department of Education, *New Jersey Student Learning Standards - Mathematics: Climate Change Companion Guide*, Trenton, NJ: New Jersey Department of Education, 2023).

2. Richard Beach, "Addressing the Challenges of Preparing Teachers to Teach about the Climate Crisis," *The Teacher Educator* 58, no. 4 (2023): 507–22. doi:10.1080/08878730.2023.2175401.

3. Lauren Madden et al., "Parents' Perspectives on Climate Change Education: A Case Study From New Jersey," *ECNU Review of Education*. doi:10.1177/20965311231200507.

4. Piers M. Forster, et al., "Indicators of Global Climate Change 2022: Annual Update of Large-scale Indicators of the State of the Climate System and Human Influence," *Earth System Science Data* 15, no. 6 (2023): 2295–327. doi:10.5194/essd-15-2295-2023.

5. National Aeronautics & Space Administration, "A Guide to Climate Change for Kids," May 15, 2024, https://climatekids.nasa.gov/kids-guide-to-climate-change/.

6. Terra Léger-Goodes et al., "How Children Make Sense of Climate Change: A Descriptive Qualitative Study of Eco-anxiety in Parent-Child Dyads," *PLoS ONE* 18, no. 4 (2023): E0284774. doi:10.1371/journal.pone.0284774.

7. Etienne Terrenoire et al., "The Contribution of Carbon Dioxide Emissions from the Aviation Sector to Future Climate Change," *Environmental Research Letters* 14, no. 8 (2019): 084019. doi:10.1088/1748-9326/ab3086.

8. James Nottingham and Bosse Larsson, *Challenging Mindset: Why a Growth Mindset Makes a Difference in Learning–and What to Do When It Doesn't* (Thousand Oaks, CA: Corwin Press, 2018).

9. Matthew T. Ballew et al., "Does Socioeconomic Status Moderate the Political Divide on Climate Change? The Roles of Education, Income, and Individualism," *Global Environmental Change* 60 (2020): 102024. doi:10.1016/j.gloenvcha.2019.102024.

10. John Spencer and A. J. Juliani, *LAUNCH: Using Design Thinking to Boost Creativity and Bring out the Maker in Every Student* (San Diego, CA: Dave Burgess Consulting, 2016).

11. Andrew Bauld, "Eco-Action: Turning Students' Climate Anxiety into Agency," *School Library Journal,* March 20, 2024, www.slj.com/story/eco-action-turning-students-climate-anxiety-into-agency#articleComment.

12. The Green Microgym, "The Green Read and Ride Bike," https://www.thegreenmicrogym.com/the-story-of-the-upcycle-eco-charger/for-schools/.

13. Morven Museum & Garden, "Festival of Trees 2023," www.morven.org/fy24/festivaloftrees.

CHAPTER 9

Using Data Visualization to Enhance Climate Change Education

Melissa Zrada, Kristin Hunter-Thomson, and Carrie Ferraro

The impacts of climate change will be felt throughout the world, yet these impacts will not be felt equally, with some populations being particularly vulnerable.[1] To address these impacts in a lasting, equitable manner requires that we support our future leaders and decision-makers (our students) in building the skills they need to think holistically and systematically about our planet.[2,3] Among these skills are the ability to analyze data from a variety of sources and the capacity to incorporate this information into their reasoning and decision-making processes.

Whether focusing on climate topics in the elementary, middle, or high school, we must provide our students with the opportunity to practice using and synthesizing data across subjects. Working with data from multiple disciplines and at different scales not only fosters the type of systems thinking mentioned above, but also increases relevance and provides an entry point for students with different interests.[4] Yet, there remains a lack of climate change education beyond STEM subjects[5] and limited opportunities for integration across subjects.[6] For more around working on interdisciplinary projects, see chapters 7, 10, and 13 in this book. Additionally, learners need opportunities to engage with the multiple scales (e.g., local, regional, global) and dimensions (e.g., comparing air temperature, demographics, and emission rates) of climate. One tricky aspect of engaging with these scales and dimensions of climate is that many are hard for learners to conceptualize in their heads or through words alone. For example:

- How do we consider the global climate system?
- How do we compare the global average with our local observations?
- How do we consider the interplay of atmospheric carbon dioxide levels and time?

- How do we compare this interplay from the last thirty years to fifteen years ago?

We wager that visuals of maps and/or line graphs popped into your mind as you read these questions. For those of us who are visually able, visual information is orders of magnitude faster and easier to process than written prose.[7] This need for processing speed and ease increases as the number of things we consider increases. Visuals can serve as an engagement point for learners to access the topic. Therefore, when helping our learners make sense of climate and climate change, phenomena that inherently involve multiple scales and dimensions, leveraging the incredible benefits of visualizing data can be critical.

Data visualizations increase the accessibility of visualizing large systems. Another benefit to utilizing them in teaching climate and climate change is that they make visible different periods in time and/or geographic locations in the world to your learners. For example, a fourth-grader in Newark, New Jersey, can view visualizations of climate impacts from their local area to Death Valley, California, and Juneau, Alaska (three locations with very different climates). An eighth-grader in Corpus Christi, Texas, can visualize the levels of ocean acidification in the Gulf of Mexico near their school and compare it with levels in the Mediterranean Sea and the Southern Ocean (two locations where they would not be able to collect their own data during an in-class investigation). A tenth-grader in Miami, Florida, can visualize the demographics of different neighborhoods around the city overlaid with current high-tide flooding areas and historical areas where home loans were made from time frames before they were born. Data visualizations can transport us in time and space to areas we have not personally been.[8] This affordance of data visualizations is not only a benefit but is in fact critically necessary when learners are building their understanding of climate change, because they physically cannot observe every aspect of climate change.

Data Across Disciplines

Across our math,[9] science,[10] and social studies[11] student-learning standards across the United States, we highlight the need for K–12 students to be able to work with data and data tools, as well as analyze data to answer questions about real-world problems. Research indicates that allowing students opportunities to engage and interact with data in multiple authentic ways supports them as they build their skills to understand and use data effectively.[12] Student-collected data can spur discussion of "data ownership" (i.e., provenance) and how data provide evidence for claims, interpretations, and conclusions.[13] In fact, students spend large amounts of their K–12 learning experiences with data they collect themselves.

However, as previously discussed, climate and climate change are topics that occur on temporal and spatial scales much larger than we can measure and observe in any given class period or even unit. To support our students' understanding of climate and climate change and to more broadly support data literacy and enhance critical thinking skills, students should have opportunities to transition from working with small, typically self-collected datasets to large, professionally-collected datasets.[14,15] These larger datasets make it possible to explore climate and climate change more fully and authentically from our classrooms. They also increase students' ability to make decisions about data quality, what visualization could best help them explore and explain their data, and how to make sense of what the data mean, all of which are key aspects of working with real-world data.[16,17]

If you are onboard with the idea that data not only can but *should* be used to teach climate and climate change, then what does that mean? First, let's start with some norm-setting on what we mean by these terms:

1. **Data**: Words, images, and numbers observed and recorded about a phenomenon or system to more deeply explore and understand it; information that is gathered by counting, measuring, or observing.[18]
2. **Data literacy**: Having the abilities and skills to use data within the context through which they are interpreted,[19,20] or more specifically, the ability to collect, organize, visualize, analyze, interpret, and share data for yourself and others to understand.[21]
3. **Data visualization**: Visual representation of data values (e.g., chart, map, plot, graph, data table, icon array, image, word cloud, diagram).[22]

With that as context, we think it is also valuable to think about what using data in our teaching could mean. We posit that there are two main ways we use data in our teaching: teaching data skills and teaching with data. While distinct, these two ways can be used simultaneously or incorporated into separate activities and have a complementary relationship to one another for our students' learning.

The first category, "teaching data skills," includes two major aspects that often come to mind when we think of having students work with data. For students to use data, they need to learn how to physically work with data (e.g., how to make a line graph, how to calculate slope, how to read a map legend). These can be thought of as the executional tasks of constructing visualizations or calculating statistics from data. Students also need to learn how to engage with data by making different data moves. We define a data move as a step or action a user takes as they work with and make sense of data. It can be done with or without technology. For example, when working with data, it is important to determine what data are useful (in other words, identify data) to explore for your question.[23] Once students have data, they likely will need to make their dataset usable by cleaning it so all values are recorded consistently across variables (e.g., data format, how missing values are recorded in the dataset, and how time values are recorded in the dataset).[24] These are "moves" we need to make when working with data but that are not about executing a specific task, like constructing visualizations or calculating statistics from a dataset. We refer to both learning how to physically work with data and data moves as "teaching data skills."

Meanwhile, making sense of data goes beyond just physically working with it and the data moves we perform in fact, data and data visualizations are often directly leveraged to teach the learning objective content (e.g., impacts of climate change on ecosystems); we consider this "integrating data into teaching." Data and data visualizations in this context are a means to the end of achieving content understanding. Therefore, students are applying their learned "data skills" to engage with data in the service of deepening their understanding of content. This use of data empowers users (students) to engage with sensemaking themselves and offers them a different way to engage with the content material.[25] Therefore, data and data visualizations are being integrated into the instructional activity similarly to how we integrate reading passages, diagrams, demonstrations, etc., into our teaching practices.

Climate and climate change datasets can be leveraged for both "teaching data skills" and "integrating data into teaching" in our classrooms regardless of subject area and grade level. For example, when "teaching data skills," we can use real-world atmospheric carbon dioxide and time data to help students learn how to make a line graph and understand how it is different from a line plot or bar chart. We can also leverage the same dataset and line graphs from the NOAA Global Monitoring Lab Mauna Loa resource, "Trends in Atmospheric Carbon Dioxide," to help students explore the fact that atmospheric carbon dioxide has continually increased since the late 1950s in Hawaii as a way to "integrate data into teaching" about climate change.

Classroom Opportunities

So, what could this actually look like? Below, we highlight a range of opportunities for using data visualizations to enhance students' learning of climate and climate change in a variety of classrooms: across subject areas (arts, literature/ELA, history/social studies, math, and science), across grade levels (elementary through high school), and across media (online and printed). We work with a wide range of in-service and pre-service educators across these subject areas and grade levels and have selected the following resources based on our experiences of what most resonates with audiences with whom we work and we observe being integrated successfully into classrooms across the country.

STEM ELEMENTARY SCHOOL OPPORTUNITIES WITH PAPER

It is critical to provide our youngest learners with opportunities to explore data and develop data skills. This section highlights two data education resources available for elementary-aged students: Tableau Data Kids[26] and NASA Next Gen STEM For Educators.[27]

Tableau Data Kids is a web resource created by Tableau, a data visualization tool used broadly in both industry and education. The website includes resources such as an Educator Kit, Activity Library, example visualizations, and blog posts. The Educator Kit provides pencil-and-paper activities for students ages six to thirteen. Resources within the kit include an Overview document, Icebreaker activity, "Introduction to Data" presentation, "Activity Guide: What are your favorite songs?" and additional resources. The material promotes data skills like collecting your own data and topics like qualitative versus quantitative data. Further, the Data Kids Activity Library section provides eight additional activities on a wide range of topics, such as gardening and movie characters.[28] The "Weather & Mood" is a great way to have students work with weather/climate-related data in new ways that are accessible to them, and you can design your own activity building on the students' approach to data.

The NASA Next Gen STEM For Educators website[29] allows teachers to explore STEM content across three grade spans (K–4). Although the site's focus is not solely data, the intuitive search feature allows you to search for resources by criteria like subject and keyword. For example, searching "data" with the audience set to "educators" and grade levels set to "K–4" yields nearly one hundred results, one of which is titled "Precipitation Towers."[30] This particular lesson promotes data literacy skills by prompting students to use manipulatives (blocks) to represent rainfall in various locations over the course of a year, which is a great way to encourage students to engage with weather and climate-related data in age-appropriate ways (see figure 9.1). A full lesson plan and supplementary materials are provided.

DISCIPLINARY EXPLORATIONS IN MIDDLE–HIGH SCHOOL COMBINING PAPER & ONLINE

Several resources are available for middle and/or high school teachers interested in integrating data into their science curriculum. One such established program is the

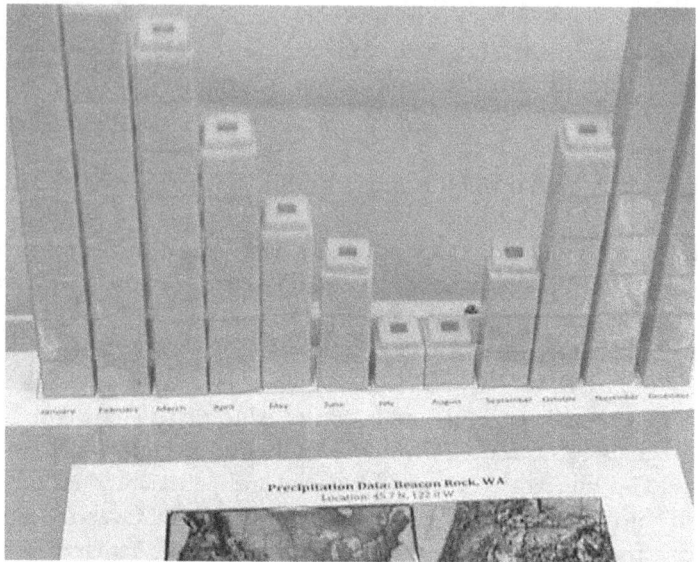

Figure 9.1   Rainfall data from NASA

GLOBE Program: "The Global Learning and Observations to Benefit the Environment (GLOBE) Program is an international science and education program that provides students and the public worldwide with the opportunity to participate in data collection and the scientific process, and contribute meaningfully to our understanding of the Earth system and global environment."[31] At the time of this writing, over 120 countries and over forty-eight thousand educators have participated in the GLOBE Program.[32] GLOBE provides resources for educators that span grades K–12 and five "protocols": Atmosphere, Biosphere, Earth as a System, Hydrosphere, and Pedosphere. Robust training opportunities are available for teachers, including tutorials and workshops. Further, there is a helpful document[33] for educators connecting the Next Generation Science Standards to GLOBE, which includes example standards and connection activities for grades K–12.

Through GLOBE, students can learn best practices for data collection, collect their own earth science data, analyze data, and share their data with others. Tools like the GLOBE Observer website or app allows students to share the data they collect with the broader scientific community, providing an exciting way for middle and high school students to benefit from student-collected data while pooling those data with a larger dataset to ask deeper questions across space and time. For example, the Urban Heat Island Effect through the Surface Temperature data collection protocol[34] is a great way for students to explore weather topics as related to local climatic experiences and think about them in a broader context (see figure 9.2).

Atmosphere Investigation

Surface Temperature Data Sheet

* Required Field

School Name: ______________________________ Study Site: ___________________________

Observer names: ____

Date: Year______ Month______ Day______ Universal Time (hour:min): _______________

**Surface Temperature*

Site's Overall Surface Condition (Select One): ❑ Wet ❑ Dry ❑ Snow

Sample	Temperature Measurement (°C)	Snow Depth (mm) (*if snow selected above)
1		❑ zero ❑ Trace (<10 mm) ❑ Measureable (>10mm) ____ mm
2		❑ zero ❑ Trace (<10 mm) ❑ Measureable (>10mm) ____ mm
3		❑ zero ❑ Trace (<10 mm) ❑ Measureable (>10mm) ____ mm
4		❑ zero ❑ Trace (<10 mm) ❑ Measureable (>10mm) ____ mm
5		❑ zero ❑ Trace (<10 mm) ❑ Measureable (>10mm) ____ mm
6		❑ zero ❑ Trace (<10 mm) ❑ Measureable (>10mm) ____ mm
7		❑ zero ❑ Trace (<10 mm) ❑ Measureable (>10mm) ____ mm
8		❑ zero ❑ Trace (<10 mm) ❑ Measureable (>10mm) ____ mm
9		❑ zero ❑ Trace (<10 mm) ❑ Measureable (>10mm) ____ mm

Comments: ____

____

**Sky Conditions (next page):*

GLOBE® 2018 Appendix - 19 Atmosphere

Figure 9.2 Atmosphere Investigation: Surface Temperature Data Sheet from GLOBE

A range of available online data visualization tools can also be leveraged to use existing and upload other datasets for students to explore and visualize. These tools are designed specifically for middle and high school students to actively interact with and explore datasets. One such example is the Common Online Data Analysis Platform (CODAP),[35] created by The Concord Consortium. Designed for grades 6–14, CODAP includes curated datasets related to climate and earth science such as Future Climate Change, Hurricanes 2005–2015, and Earthquakes. Additional datasets are available to explore, spanning both STEM and non-STEM disciplines. While CODAP provides these exemplar datasets, it is also easy to import your own data into the CODAP web portal. The For Educators: Teaching with CODAP webpage provides helpful resources such as Getting Started guides and activities for students.

MIDDLE AND HIGH SCHOOL EDUCATORS CROSS-CURRICULAR CONNECTIONS USING PAPER

As mentioned above, a dynamic problem like climate change requires system wide solutions. To discover these solutions, our students must understand and apply concepts from multiple subjects, which also necessitates teachers having purposeful opportunities to discuss and co-plan strategic, meaningful, and purposeful connections across courses.

For example, in 2023, given the recent statewide requirement to integrate climate change across content areas, New Jersey educators from across subject areas and grade bands gathered with Rutgers University subject-area experts to brainstorm ways to collaborate across subject matters and institutions. To help participants get in the mindset of multi-subject coordination, meeting organizers developed a new activity utilizing datasets (lines of evidence) from multiple perspectives and disciplines. Participants were tasked with reading, interpreting, and discussing four lines of evidence demonstrating the causes and effects of climate change in New Jersey from the perspective of the arts, literature, history, and science. These datasets included maps of redlining in Newark (see figure 9.3), art installations detailing precipitation rates (see figure 9.4), first-person accounts from a recent flooding event, and graphs depicting temperature and precipitation over time.

Participants were asked to discuss with each other ways the evidence was connected and how they could use data like that provided in their classrooms. Through these conversations, educators brainstormed how data in various forms can be used as a means of integrating climate change into the curriculum across subject areas. The following list represents some of the big ideas derived from these conversations.

USING DATA VISUALIZATION 107

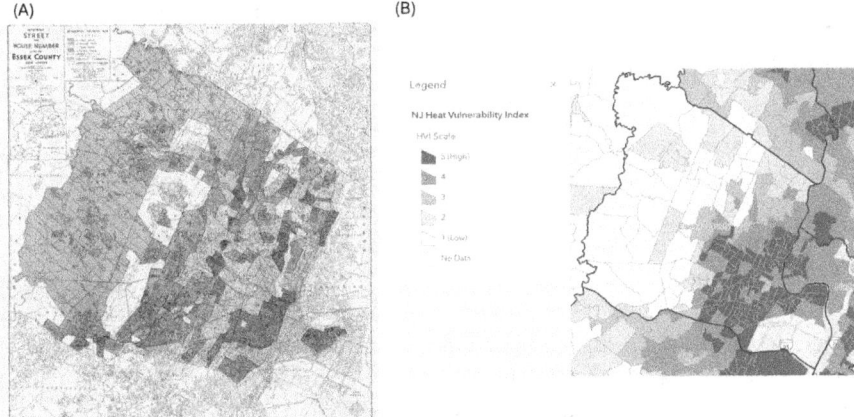

Figure 9.3 Two maps of Essex County, NJ. The map on the left (A), which was obtained from the National Archives' City Survey Files, 1935-1940, is from 1939. This map demonstrates redlining in the county through which red areas tended to have larger immigrant and black populations and green or blue area had larger white populations. The map on the right (B), obtained from NJ Public Health Adapt and created using data from 2015-2019, shows the area with high heat vulnerability in blue moving down to those with lower vulnerability in yellow. Educators were asked to compare maps like these and look for patterns.

Figure 9.4 This artwork by Adriane Colburn, commissioned by the National Oceanographic & Atmospheric Administration and the NJ Department of Environmental Protection, uses sculpture to convey streamflow data from the late 1880s to 2020 in an area of Passaic, NJ. The larger the circle is in diameter, the wetter the year. The color of the circle represents the average temperature for that year, with red being the warmest and blue being the coldest numbers recorded over this time period. *Source*: https://www.adrianecolburn.com/new-gallery-5

1. Climate change content should be presented at both local and global scales across subject matter.
2. Art and narrative data can generate interest and be used as an entry point into the topic.
3. There is a large opportunity to problematize climate change in the classroom and utilize data from different sources to find solutions.

Ultimately, the discussions highlighted the value of incorporating non-STEM data into climate change teaching in increasing interest and fostering systems thinking skills.[36] Building on these initial conversations, participants and others across the state engaged in strategic brainstorming conversations of where and how to leverage connections across subject areas around data and climate change in the curriculum (see figures 9.5 and 9.6).

Recommendations for Leveraging Data Visualizations and Climate

These classroom opportunities represent just a small handful of ways you can use data visualizations to enhance your teaching of climate and/or climate change topics. We recognize that the curricula we use, tools/platforms to which we have access, and scope and sequence of learning varies greatly across different schools and districts, so we also want to share some lessons learned from our experiences of utilizing climate data and data visualizations with students. Below, we outline four areas we pay attention to when planning instruction and reviewing instructional materials. This is not an exhaustive list, but is intended to provide a starting point for things to consider.

First, students often look at data and data visualizations that only have one to two variables. While it is important for younger learners to look at fewer variables at one time, as discussed, climate and climate change are system wide phenomena. Therefore, to help our students engage with these topics, it is important to open the kinds of data and data visualizations with which our students work. What could that look like? At the elementary level, when our students are not ready to look at multiple variables on one graph (in other words, no line graphs or scatterplots), have students compare two side-by-side line plots or bar charts. Providing the extra graph in this learning experience plants the seed that we are looking at a complex system that cannot be fully explained with one variable or visualization alone. This sets a critical foundation to build upon in middle school. In middle school, we recommend introducing students to datasets that include more than two variables. Even adding a third into the dataset prompts students to make choices about what data to look at for different questions and

USING DATA VISUALIZATION 109

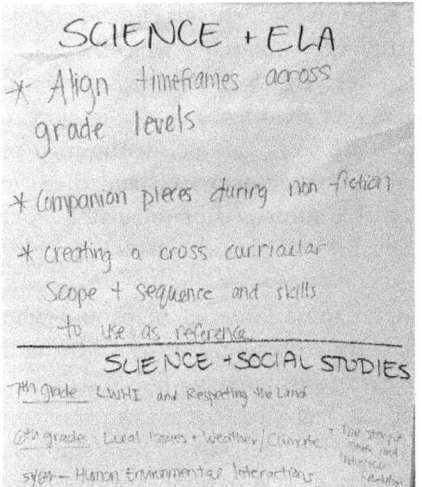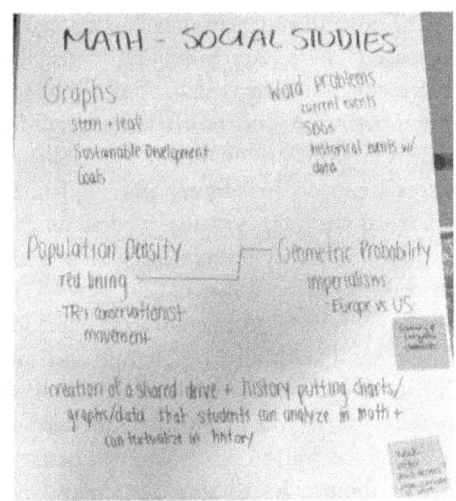

Figure 9.5. and 9.6. Examples of cross disciplinary brainstorming sessions for how to collaborate across subject areas around climate change topics using data at the middle school (a) and high school (b) levels within a district.

perspectives. In high school, students can build upon this by pooling multiple datasets together. Each of these steps creates opportunities for students to engage with the data and data visualizations for a complex topic to push their understanding in developmentally appropriate ways.

Second, *many* climate and climate change data visualizations are available online these days. We recommend a two-pronged approach for how to make choices about which to use, when, and how. One approach, select data visualization types with which students are already familiar. In table 9.1, we outline when common data visualization types are developmentally appropriate to introduce to K–12 students. Look for data visualizations from previous grade levels if you want to "integrate data into teaching" or at your grade level(s) if you want to "teach data skills".[37] Whether using a visualization type with which students are familiar, simply learning about visualizations, or seeing one for the first time, it can be extremely valuable to spend time helping students orient to the visualization type before asking them to make sense of the data.[38] This can be student-led or teacher-facilitated, but the key is creating time for students to digest what they are looking at on the page/screen before we ask them to start noticing, wondering, analyzing, or interpreting the data. You can use just one or both approaches throughout your teaching around climate and climate change; the key is going with what feels most comfortable for you and your students.

Third, in our experience, a key part of orienting students to climate and/or climate change data is explicitly ensuring students understand what the colors represent in a data visualization. While this is relevant to all data visualizations, the time frames and/or broad geographic areas that are often included in climate data visualizations makes it especially important. For example, in the map of Essex County, (see figure 9.4), it is critical that students understand that the color on the map visually represents the variable measured or calculated (e.g., heat vulnerability index), not that the land in those neighborhoods is that color in real life.

As adults who are comfortable looking at these kinds of maps and who understand both climate change and the data we are looking at, it can be hard for us to remember our learners often take data maps literally. This disconnect between how you and your students perceive a data visualization can result in a lot of frustration of them "not getting it." Taking the time to help students orient themselves to this particular aspect of data visualizations can pay huge dividends in the long run when integrating data into teaching climate and/or climate change.

Finally, given that climate and climate change are about complex, large-scale systems, it is critically important to include statistical thinking moves into all sensemaking around these datasets and data visualizations. Our first recommendation of expanding dataset sizes and the number of data visualizations students work with starts to get at this issue, but we recommend integrating more elements into your teaching practices, regardless of your learners' ages or the subject area you teach. We do not advocate that all students learn how to make statistical calculations; rather, whenever we work with data, we need to think statistically about them.[39] What does that mean? Whenever we have data, we are working with a smaller part of all that could possibly be measured of the system or phenomenon. Helping our students think about any dataset as a part of the larger whole is critical for working with data. Here are four ways we suggest helping students learn to think about datasets and data visualizations as parts of the whole:

1. Explicitly discuss and consider the sample size for data you have.
2. Discuss the variability in data as a natural component of the real world (not a sign that something is wrong with the data).
3. Frame discussions of data and data visualizations around supporting or refuting a claim with evidence from that data, not *proving* things with data.
4. Assist students in making a claim or conclusion from the data itself before having them make an inference or prediction about the larger system or phenomenon the data supports.

This is not an exhaustive list of how we can help our students think statistically about data, but in our experience, these recommendations provide a great starting point for all levels of learners to build a mindset of how to think more broadly and statistically about data. When working with climate and climate change data, it is especially important to consider the sample, natural variability, supporting evidence, and the difference between claims and predictions. The climate system is large and complex, so considering these four aspects helps students make sense of the data and understand why experts share the specific claims they make from the data, rather than wider claims you may anticipate.

We already pay attention to many things when planning our instruction and reviewing instructional materials. With the extra layers of climate and/or climate change being complex systems and data skills being critical twenty-first-century skills, juggling it all can start to feel really overwhelming. Therefore, we suggest choosing one or two of these four recommendations as a starting point:

1. Have students look at more than one visualization of the same dataset and/or more variables in a dataset than they need for any one question.
2. Select existing climate-related data visualizations of a type students have previously mastered if you want to "integrate data into teaching" or those specific to a grade level if you want to "teach data skills" through it.
3. Explicitly ensure that students understand what is being represented by each part of a data visualization (e.g., color).
4. Include developmentally appropriate ways for students to make or consider statistical thinking moves in their sensemaking of the datasets and/or data visualizations.

Conclusion

It is tricky to get our students to make sense of climate and climate change because it inherently involves multiple scales and dimensions. Helping students make sense of data and data visualizations can also be tricky. However, leveraging data visualizations can be rewarding and is critical to helping learners make sense of the climate and/or climate change. In fact, working with real-world climate data and visualizations in different ways throughout the process of making sense of these phenomena is a critical skill for students' success with the topic and in their broader lives, given the pervasiveness of climate-related data visualizations.

There are so many ways we can and are getting students to use data visualizations as they explore climate and/or climate change. We have shared some of our favorite ways. What strategies are you using to leverage data as a tool to enhance how your students explore and make sense of climate and/or climate change topics? What could you now try?

Glossary

Attribute: A characteristic or property of an object or event that is observed and/or measured. For example, if we are interested in looking at our classes' reading habits, examples of attributes to collect data on could include favorite genre, number of pages read per week, number of books read, etc. Each is an attribute about what and when each of us read to make up ways to explore our classroom reading habits.

Bar graph/chart: A visualization with horizontal or vertical bars that represent categories of data. The length of each bar generally represents the frequency or average of data values of the category.

Case: The group of measurements or observations of different attributes about the same object or event from one instance of measurement. For example, a case could include a person's shoe size, their eye color, and their favorite ice cream flavor. In this example, the case is the person.

Data: Words, images, or numbers observed and recorded about a phenomenon or system to more deeply explore and understand it; information gathered by counting, measuring, or observing.

Dataset: An organized collection of data values for each attribute or variable measured for a case.

Data literacy: Having the abilities and skills to use data within the context in which the data are interpreted; more specifically, the ability to collect, organize, visualize, analyze, interpret, and share data for yourself and others to understand.

Data move: A step or action a user takes as they work with and make sense of data.

Data visualization: Visual representation of data values (e.g., chart, map, plot, graph, data table, icon array, image, word cloud, diagram).

Line plot: A graph that displays numerical data for each case as a point or an icon above a number line. The line plot represents the frequency of all data values for a particular attribute.

Line graph/chart: A visualization that uses lines to connect plotted data values that are ordered in nature (e.g., May 3 always follows May 2 and precedes May 4, so dates are ordered). The line represents likely data values if they had been collected.

Map: A visualization that represents the geospatial location of data.

Represent: To show, symbolize, or stand for something.

Variable: A characteristic or property of an object that is measured; synonym for *attribute*; often used to refer to attributes in some subject areas.

Notes

1. Hoesung Lee et al., *Climate Change 2023: Synthesis Report. Contribution of Working Groups I, II and III to the Sixth Assessment Report of the Intergovernmental Panel on Climate Change.* (Canberra, Australia: Australian National University, 2023).
2. Liesl Hotaling et al., "Educating with Data." In *Exemplary Practices in Marine Science Education: A Resource for Practitioners and Researchers*, ed. Geraldine Fauville et al. (New York, NY: Springer International, 2019): 207–23.
3. US Global Change Research Program, *Global Climate Change Impacts in the United States* (Cambridge, UK: Cambridge University Press, 2009).
4. Deana Pennington et al., "Bridging Sustainability Science, Earth Science, and Data Science through Interdisciplinary Education," *Sustainability Science* 15 (2020): 647–61.
5. Richard Beach, "Addressing the Challenges of Preparing Teachers to Teach about the Climate Crisis," *The Teacher Educator* 58, no. 4 (2023): 507–22.
6. Efrat Eilam, "Climate Change Education: The Problem with Walking Away from Disciplines," *Studies in Science Education* 58, no. 2 (2022): 231–64.
7. Jonathan Schwabish, *Better Data Visualizations: A Guide for Scholars, Researchers, and Wonks* (New York, NY: Columbia University Press, 2021).
8. R.J. Andrews, *Info We Trust: How to Inspire the World with Data* (Hoboken, NJ: John Wiley & Sons, 2019).
9. Common Core State Standards Initiative, *Common Core State Standards for Mathematics* (Washington, DC: Common Core State Standards Initiative, 2010).
10. National Research Council, Next Generation Science Standards: For States, By States (Washington, DC: National Research Council, 2013).
11. National Council for the Social Studies, *The National Standards for the Preparation of Social Studies Teachers* (Silver Spring, MD: NCSS, 2018).
12. Tim Erickson et al., "Data Moves," *Technology Innovations in Statistics Education* 12, no. 1 (2019).
13. Barbara Hug and Katherine L. McNeill, "Use of First-Hand and Second-Hand Data in Science: Does Data Type Influence Classroom Conversations?" *International Journal of Science Education* 30, no. 13 (2008): 1725–51.
14. Kim Kastens, Ruth Krumhansl, and Irene Baker, "Thinking Big," *The Science Teacher* 82, no. 5 (2015): 25.
15. Elizabeth H. Schultheis and Melissa K. Kjelvik, "Using Messy, Authentic Data to Promote Data Literacy & Reveal the Nature of Science," *The American Biology Teacher* 82, no. 7 (2020): 439–46.
16. Kirstin Fontichiaro and Jo Angela Oehrli, "Why Data Literacy Matters," *Knowledge Quest* 44, no. 5 (2016): 21–27.
17. Kim Kastens and Margie Turrin, *Earth Science Puzzles: Making Meaning from Data* (Richmond, VA: NSTA Press, 2010).
18. Andrews, *Info We Trust*.
19. Erica Deahl, "Better the Data You Know: Developing Youth Data Literacy in Schools and Informal Learning Environments," M.S. thesis (SSRN, 2014).

20. Chantel Ridsdale et al., *Strategies and Best Practices for Data Literacy Education: Knowledge Synthesis Report* (January 2015).

21. Kristin Hunter-Thomson, "Data Literacy 101: How Do We Set Up Graphs in Science?" *Science Scope* 42, no. 2 (2018): 78–82.

22. Tableau, "Tableau Data Kids," https://www.tableau.com/academic/data-for-kids.

23. Michelle Wilkerson et al., "Data Moves: Restructuring Data for Inquiry in a Simulation and Data Analysis Environment," in *Rethinking Learning in the Digital Age: Making the Learning Sciences Count*, 13th International Conference of the Learning Sciences (London, UK: ISLS, 2018).

24. Erickson et al., *Data Moves*.

25. Andrews, *Info We Trust*.

26. Tableau, "Tableau Data Kids."

27. NASA, "Next Gen STEM For Educators," https://www.nasa.gov/learning-resources/for-educators.

28. Tableau, "Tableau Data Kids."

29. NASA, "Next Gen STEM For Educators."

30. NASA, "Precipitation Towers," https://www.nasa.gov/stem-content/precipitation-towers/

31. https://www.globe.gov/.

32. The GLOBE Program, "GLOBE Countries and Members Map," https://www.globe.gov/globe-community/community-map.

33. The GLOBE Program & NGSS, "The Globe Program and NGSS: A Tour through Elementary, Middle and High School Grades," https://www.globe.gov/documents/10157/9e9875cc-fd39-4c92-8419-861937a93bb5

34. The GLOBE Program, "Urban Heat Island Effect Surface Temperature Intensive Observation Period," https://www.globe.gov/web/surface-temperature-field-campaign

35. https://codap.concord.org.

36. James Shope et al., in prep.

37. Andrews, *Info We Trust*. https://infowetrust.com/

38. Hotaling et al., "Educating with Data." https://link.springer.com/chapter/10.1007/978-3-319-90778-9_13

39. Kristin Hunter-Thomson, "Why Should We All Embrace Statistical Thinking?" *Science Scope* 46, no. 3 (2023): 6–11.

CHAPTER 10

Socioscientific Issues as a Framework for Teaching Climate Change

Sami Kahn and Timothy Lintner

Climate change presents a rich curricular context for critical exploration in both science and social studies education. However, science and social studies are often siloed in K–12 teaching, making it difficult to provide meaningful real-world experiences for students that prepare them for informed and engaged citizenship. The Socioscientific Issues (SSI) framework capitalizes upon the overlaps between science and social studies education while promoting moral development and inclusive education practices. This chapter will explore the use of the SSI framework to teach climate change across multiple content areas, including science and social studies. After introducing the SSI framework and research supporting its use for promoting inclusive science and social studies literacy, we will present a flexible lesson plan that can be incorporated into any unit on climate change over a range of grade levels, as well as recommendations for best practices for implementing SSI in the classroom.

Goals of Science Education: Scientific Literacy

To appreciate the value of climate change education for science and social studies education, it helps to understand a bit about each field's underlying discipline and goals. Science is the systematic study of the natural world through observation and experimentation. Therefore, science is empirical, as it relies on collecting data to understand the natural world. Scientists observe phenomena such as sea-level rise or polar ice cap loss, develop testable questions about these phenomena, formulate hypotheses, engage in data collection and analysis, and develop conclusions based on their findings. While science in school is often taught as a singular "scientific method," in reality, scientists may take different routes to gain knowledge. Therefore, it is critical to help students understand that science is more than following a recipe; it is a creative and nimble way of understanding the natural world. Science is also a collaborative process, as scientists and engineers often work in teams and engage in discourse to advance their respective fields by publishing and presenting findings or simply through direct communications among labs. In this way, the scientific endeavor is quite different from cartoon depictions of solitary scientists having "eureka" moments in a lab. Therefore, developing future scientists who can clearly communicate their work, engage in respectful discourse about it, and appreciate different perspectives and approaches is key to the success of scientific advancement.

A key goal of science education is referred to as scientific literacy.[1] While early conceptions of scientific literacy involved memorizing facts and conducting labs that essentially confirmed predetermined outcomes, contemporary views of scientific literacy emphasize all students' ability to apply scientific knowledge to everyday life; that is, students should be able to use science for decision making in their personal lives, in their work, and as informed and engaged citizens. In this view, science education is inclusive and empowering because scientifically literate individuals can evaluate evidence and make informed decisions on everything from which toothpaste to use to what climate policies to support. To promote students' meaningful application of science learning, contextualizing science content within societal issues typically situated in social studies education can help them recognize the importance of science in societal decision making and hopefully appreciate and enjoy both science and social studies even more!

Goals of Social Studies Education: Social Studies Literacy

Just as science is the systematic gathering of information, so, too, is social studies. Though mired in the "traditional" perception that social studies is nothing more than the mere collection—and compartmentalization—of names, dates, places, and faces, social studies teaching and learning is rooted in inquiry. Here, students generate questions based on what they read or observe. These initial questions ask students to use disciplinary tools (e.g., research, interviews, maps and models, etc.) to contextualize their questions. To put it simply, students gain a deeper, more layered, complex, and connected understanding of issues of interest.

Next, students gather and evaluate information specifically looking for collaboration, bias, and/or validity. The final step in the inquiry process, and arguably the most important, asks students to communicate their conclusions and take civic action. This four-step inquiry arc replaces the often-passive mode of social studies teaching and learning by creating active and responsible student-citizens who can identify and analyze public problems, respectfully and constructively deliberate with others about important issues, and take constructive action that influences civic decisions large and small.

This inquiry-based model of instruction is the cornerstone of social studies literacy. More than just noting a numeric rise in global temperatures, we—in both science and social studies classrooms—need to ask deeper, substantive, and discipline-connected questions. These questions spur interest, emotion, and action. Both science and social studies ask students to scratch their heads, do the requisite research, then raise their hand, ask the tough questions, and challenge the answers given. Whether historical detectives or eco-warriors, the ultimate goal of science and social studies education is the same: to encourage our students to become informed citizens unafraid to take civic action. After all, it's their world. too!

Standards Related to Climate Change in NGSS and NJSLS-S

The New Jersey Student Learning Standards for Science (NJSLS-S)[2] and the Next Generation Science Standards (NGSS)[3] on which they are based both attempt to balance students' competence in science content knowledge and the application of that knowledge in order to support an educated and competitive workforce and develop informed and engaged citizens. The NGSS and NJSLS-S provide teachers and schools with a schema for promoting and assessing scientific literacy for students in K–12 classrooms.

The standards are divided into three disciplinary areas: life science, physical science, and earth and space science. The standards in both documents are written as "performance expectations" that state exactly what proficient students should be able to do to demonstrate their knowledge. Each standard is also built upon three major dimensions: scientific and engineering practices, crosscutting concepts that unify the study of science and engineering, and core ideas in the major disciplines of natural science. Finally, both sets of standards are aligned with the Common Core standards for ELA/literacy and math. Even with all that information, neither the NGSS nor the NJSLS-S provide specific curriculum for teachers and schools; rather, they allow a great deal of leeway for teachers to flex their creativity and explore the wide range of resources available.

Standards Related to Climate Change Education in NCSS, C3, and NJSLS-SS

The National Council for the Social Studies (NCSS)[4] outlines ten thematic strands (or standards) that undergird powerful social studies instruction. Though climate change is not explicitly mentioned, there does exist tangential yet supportive correlations between science, social studies, and climate change. *People, Places and Environments* asks students to investigate the impact of human activities on the environment; *Civic Ideas and Practices* challenges students to take civic action by questioning local, regional, national, or global policies and practices; and *Science, Technology, and Society* encourages the use of empirically driven scientific outcomes to influence civic perspective and policy.

To offer additional guidance to K–12 social studies educators, NCSS produced the College, Career, and Civic Life (C3) Framework for Social Studies State Standards.[5] Here, the focus is less on factual representations of social studies content and more on the development and employment of inquiry-based critical thinking skills that lead to participatory thought and action. This framework, or the aforementioned inquiry arc, situates social studies as a series of sustained, structured opportunities for students to grapple with content and draw informed, reasoned conclusions that result in civic engagement and action.

Blending NCSS's thematic strands or topical coverage with the inquiry/action focus of the C3 framework rests on the New Jersey Student Learning Standards—Social Studies (NJSLS-SS). Climate change is explicitly addressed in the NJSLS-SS, yet is rooted in inquiry, discovery, evaluation, and action. Like the C3 framework, the NJSLS-SS grounds instruction in seven pillars of inquiry-based teaching and learning: developing questions and planning inquiry, gathering and evaluating sources, seeking diverse perspectives, developing claims and using evidence, presenting arguments and explanations, engaging in civil discourse and critiquing conclusions, and taking informed action. Common to NCSS, C3, and NJSLS-SS is the instructional latitude these frameworks provide K–12 educators in designing and delivering sustained opportunities for students to become civically engaged with issues that impact their lives.

In 2020, New Jersey became the first state in the nation to adopt standards in seven out of nine content areas that support climate change education, including science and social studies. The standards that support climate change learning[6] are incorporated into the NJSLS, and teachers are encouraged to design interdisciplinary climate change lessons/units to address multiple content area standards. We have found the SSI framework to be a formidable approach to addressing the science and social studies climate change standards in a cohesive, engaging, and empowering manner!

Socioscientific Issues Framework: Theory and Practice

SSI is a framework in which students investigate complex, often controversial societal issues related to science as a way of promoting scientific literacy and engaged citizenship.[7] Typical SSI topics might include animal testing, water fluoridation, coastal remediation, drone policies, and genetic engineering, among others. Of particular importance is that decision making around SSI is informed by, but not completely answered by, science in that SSIs necessarily involve ethical dilemmas or "should" questions that touch on aspects of politics, economics, philosophy, and so on. So, while one could determine, for example, that offshore wind farms are an efficient and effective approach to providing electricity in New Jersey (a "scientific" determination), one must also consider the economic, political, aesthetic, and philosophical implications for a range of stakeholders, such as fisherman, local residents, the tourism industry, environmental organizations, and so on (therefore making an "ethical" determination). To negotiate such complex issues in the classroom, students engage in discourse and debate using evidence-based argumentation to support their claims. A strong research base suggests that students who participate in SSI activities improve their ability to evaluate sources and quality of evidence, retain science content, apply science to real-world contexts, develop and communicate arguments, take others' perspectives, and appreciate the consequences of their actions (or inactions) on others.[8]

It is particularly notable that engagement in SSI promotes students' moral development; that is, by gaining an appreciation of others' perspectives and recognizing the consequences of their actions, students become more empathic, open-minded, and world-centric rather than egocentric. This is perhaps the most persuasive reason for using SSI in the classroom, particularly for climate change issues, as such issues require deep engagement with weighing diverse perspectives around environment, economy, politics, and personal philosophy. Students must be prepared to see themselves as voices of change while being able to integrate others' voices in an informed, thoughtful, evidence-based way, rather than simply becoming more entrenched in their own viewpoints.

In the next section, we provide a model lesson plan to demonstrate the power of SSI as applied to a controversial question around climate change.

Model Lesson Plan Vignette: A United Nations Summit on Climate Change Priorities

The lesson we present positions students as participants in a United Nations Climate Summit. To prepare for the summit, students must research various climate change priorities and approaches to mitigation, adaptation, and resilience; weigh competing arguments; and develop their priorities in climate change decision making. This lesson can be adapted for students across K–12 levels; however, for ease of presenting the standards alignments, we present the lesson at the middle school (eighth-grade) level. We are confident that educators, as experts in their curriculum and students' developmental stages, will be able to adapt the lesson accordingly. Nonetheless, we provide some tips for adapting the lesson for younger and older students at the end of the lesson. As the lesson draws together standards from both science and social studies, we present it within a vignette of two collaborating teachers, Ms. Vasquez, a science teacher, and Mr. Wiggins, a social studies teacher. The New Jersey Department of Education strongly supports the development of interdisciplinary lessons and units to address climate change education, and we hope this lesson amplifies the importance of such an approach.

Ms. Vasquez and Mr. Wiggins were in an eighth-grade-level meeting when the topic of climate change education came up. They both knew climate change education was a new requirement across the curriculum in New Jersey and were eager to address the standards in their subject areas. However, they were also a bit nervous about it, because they were concerned that developing a new curriculum and finding time to implement it in the existing schedule would be tricky. After chatting a while, Ms. Vasquez mentioned she had read about a UN Summit on Climate Change and thought it would be interesting to have students simulate one themselves. She thought it would be exciting for students to investigate and prioritize different actions around climate change, given the fact that her students often hear things they should do about climate change, like taking public transportation and riding bicycles rather than taking cars, but that they really didn't have a good sense of the various contributors to climate change or the range of remedies available to address it.

Mr. Wiggins was intrigued by the idea of the UN Summit, as his teaching practice is deeply rooted in the idea that social studies education should prepare students with the knowledge and skills to improve their local and global communities through civic action and discourse. Mr. Wiggins envisioned evidence-based argumentation as the basis for a climate change summit simulation, prompting students to ask questions, conduct research, assess the quality of evidence, weigh competing perspectives, and perform other key practices in social studies education. By the end of the meeting, Ms. Vasquez and Mr. Wiggins decided to collaborate on an interdisciplinary lesson. They met again to look at various resources on the topic, including the NJ Climate Education Hub[9] and the UN Climate Change education website.[10] The lesson they developed looked like this:

SUGGESTED GRADE LEVEL(S): 8TH

Driving question: How should we prioritize efforts to combat climate change?

Lesson overview: Students are introduced to the ways climate change impacts are felt across different sectors (e.g., energy, food, water, land, biodiversity). Students then determine the sector they feel should be prioritized and identify approaches to combat climate change effects for that sector.

Ethical issues: Allocation of resources, human-environment interactions, stewardship, public good/personal responsibility, government vs. private sector, wildlife conservation

Connecting to NJSLS-S and NJSLS-SS

ESS3.C: HUMAN IMPACTS ON EARTH SYSTEMS

- MS-ESS3-3: Apply scientific principles to design a method for monitoring and minimizing a human impact on the environment.[11]
- MS-ESS3-4: Construct an argument supported by evidence for how increases in human population and per-capita consumption of natural resources impact Earth's systems.[12]

ESS3.D: GLOBAL CLIMATE CHANGE

- MS-ESS3-5: Ask questions to clarify evidence of the factors that have caused climate change over the past century.[13]

MS-LS2: ECOSYSTEMS: INTERACTIONS, ENERGY, AND DYNAMICS

- MS-LS2-4: Construct an argument supported by empirical evidence that changes to physical or biological components of an ecosystem affect populations.[14]
- MS-LS2-5: Evaluate competing design solutions for maintaining biodiversity and ecosystem services.[15]

CIVICS, GOVERNMENT, AND HUMAN RIGHTS: PARTICIPATION AND DELIBERATION

- 6.3.8.CivicsPD.1: Deliberate on a public issue affecting an upcoming election, consider opposing arguments, and develop a reasoned conclusion.

- 6.3.8.CivicsPD.2: Propose and defend a position regarding a public policy issue at the appropriate local, state, or national level. Members of society have the obligation to become informed of the facts regarding public issues and to engage in honest, mutually respectful discourse to advance public policy solutions.
- 6.3.8.CivicsPD.3: Construct a claim as to why it is important for democracy that individuals are informed by facts, aware of diverse viewpoints, and willing to take action on public issues.[16]

CIVICS, GOVERNMENT, AND HUMAN RIGHTS: PROCESSES, RULES, AND LAWS

- 6.3.8.CivicsPR.4: Use evidence and quantitative data to propose or defend a public policy related to climate change.
- 6.3.8.CivicsPR.5: Engage in simulated democratic processes (e.g., legislative hearings, judicial proceedings, elections) to understand how conflicting points of view are addressed in a democratic society.[17]

Ms. Vasquez and Mr. Wiggins anticipate this lesson will take multiple class periods and can be divided between their two classrooms given that they share the same students. They decide on the following timing and sequence using the 5E Instructional Model[18]:

- Day 1: Engage. In science, students watch a short Khan Academy video on climate change[19] and discuss their initial thoughts on how to prioritize solutions to climate change impacts through a "four corners" discussion. In social studies, students explore human-environment interactions by characterizing facets of the relationship between their school and the environment as dependence, adaptation, or modification.
- Day 2: Explore. In science, students conduct online research on climate change impacts on various sectors such as water, food, biodiversity, etc., and record their findings. In social studies, students review the foundational concepts of the Bill of Rights, specifically the First Amendment, and investigate ways people have used their rights of speech and assembly to advocate for change.
- Day 3: Explain. In science, after sharing their findings, students are introduced to approaches to combating climate change, including mitigation, adaptation, and resilience. In social studies, students create a climate-change advocacy statement in their choice of medium.
- Day 4: Elaborate. In science, students prepare for a mock UN Climate Change Summit where they will argue for their position on which sector

should be prioritized and in what ways the impacts on that sector should be addressed. In social studies, students participate in the mock summit and teams are assessed for their arguments.
- Day 5: Evaluate. In both science and social studies, students develop and submit an individual position paper outlining their own priorities, evidence for their conclusions, and a climate change action plan.

Ms. Vasquez begins the first day's lesson (Engage) in science class by showing a short video[20] she found on the NJ Climate Change Education Hub on climate change causes, impacts, and solutions. She then reviews the video with her students by asking about the causes of climate change, including increased CO_2 in the atmosphere from a range of sources. Ms. Vasquez then asks the class, "Why is climate change a problem?" Her students share various responses such as, "It causes flooding and sea level rise," "It is harmful for some animals/plants," "It makes the world too hot," etc. Ms. Vasquez then posts five signs around the room that read as follows: 1) Energy: shifting to renewable energy, 2) Food: encouraging responsible eating and farming, 3) Water: protecting earth's freshwater and ocean resources, 4) Land: using land responsibly, and 5) Biodiversity: protecting plants and wildlife. She then asks her students, "If you were able to work on solving only one challenge at a time, which would it be and why?" Knowing the students might be inclined to go to the signs their friends choose, Ms. Vasquez asks them to jot down their answers and reasons in their notebooks, on their iPads, or on paper. She then instructs them to go stand by the sign they feel represents the most important climate impact. When the students get to their areas, she asks them to take a few minutes to discuss their thoughts on why they chose that area. (If a student is alone in an area, the teacher can join the student and discuss their position with them). After a few minutes, Ms. Vasquez asks the students to choose a spokesperson who can share the group's thoughts with the entire class.

After groups report on their discussions, Ms. Vasquez asks, "Did anything surprise you when you heard other people's ideas?" She elicits all ideas and finds that several students noted their surprise that some solutions that hadn't seemed important to them before now seem more important. She then asks, "Does anyone have a different opinion on this question now that they've heard other points of view?" In this way, she is raising the idea that it is okay for students to reconsider their ideas and remain open to different perspectives. Several students indicate they are open to having different opinions on this question. She asks, "If you have a different opinion now, what types of arguments were particularly persuasive to you?" This question helps students begin to think about the persuasive quality of evidence, particularly if they presented facts with which they were familiar. Importantly, Ms. Vasquez asks her students, "Do you feel you have enough information on this topic to make a 'final' decision on it?" Several students indicate they're not sure about land use and that they don't know what biodiversity has to do with climate change. She commends her students on their honest assessment of their knowledge and shares that they're going to have the opportunity to learn more about each of these solutions during the week. To get started, she asks, "What other information would you like to know to make an informed decision?" In this way, she is helping students develop questions and take an active role in their learning.

Later that day in social studies, Mr. Wiggins introduces students to one of the five themes of geography, human-environment interaction. He asks, "Think about this school. Before it was built, what do you think was here?" Students generate responses ranging from trees to bushes or brush to certain animals and insects. He then asks, "Given that all this stuff was here, was it worth displacing all of it to build a school?" Mr. Wiggins shows various images of how humans have impacted (both positively and negatively) their environments (e.g., freeways, strip mines, parks, shopping malls, airports, deforestation, dams, etc.). He then explains that there are three types of human-environment interactions: dependence, adaptation, and modification. Mr. Wiggins divides the class into three groups, with each being assigned to dependence, adaptation, or modification. Each group is given a large sheet of butcher paper. Students are asked to generate ideas about how their school is either dependent upon; has adapted from, with, or to; or modified the environment. After ten minutes, each group presents their conclusions to the class. Mr. Wiggins closes the lesson by stating that for the next few days, the class will explore how human-environment interaction has influenced climate change and, most importantly, what they, the students, can do about it.

The next day (Explore), Ms. Vasquez has students work in teams to research each of the different sectors or areas for climate change action (e.g., energy, food, water, land, and biodiversity), each team is challenged to answer a series of questions for each sector/solution that contributes to or is impacted by climate change: 1) What role does this sector play in climate change? 2) What are the impacts of climate change on this sector? 3) How have increases in human population or per capita (person) consumption of natural resources impacted this sector? 4) What data/evidence supports these claims? Ms. Vasquez provides teams with laptops/iPads; information for the UN Climate Action website, Explainers: Transforming climate issues into action;[21] and graphic organizers listing each of the sectors and including columns for "Role in climate change," "Impacts of climate change on this sector," "Impact of increased human population/consumption on this sector," and "Data/evidence to support this claim." The graphic organizer also has a section for "Solutions to combat climate change" for each sector, but students are instructed not to complete it yet. Students spend the remainder of this class working on their research as Ms. Vasquez monitors work, answers questions, and assists as needed.

That afternoon in social studies, Mr. Wiggins reintroduces the Bill of Rights, specifically focusing on the First Amendment. Mr. Wiggins states, "I think we need to raise the voting age to 21 because teenagers don't vote anyway." He pauses and then says, "Do I have the right to say this, even if you don't agree with me?" He spends time going over the basic tenets of the First Amendment, reminding students that as citizens, they have the right to express their ideas through speech (what they say) and the press (what they write). They can get together with like-minded individuals (assemble) and ask the government to change policies or practices (protest). Mr. Wiggins distributes laptops to each student with the charge that they are to "stroll through history" researching the ways individuals or groups of citizens have advocated for change. Students will note the name of the individual or group, the year, the issue or problem they confronted, and the outcome or resolution. Students will be given twenty minutes to work on this individual research. With ten minutes remaining, Mr. Wiggins will ask a few students to share their findings. He will close by asking, "Do you think advocating for change is worth it? What made that person or that group of people stand up and advocate for change?" He will also ask the rhetorical question, "When was the last time *you* stood up and advocated for change?"

In the next science class (Explain), Ms. Vasquez has students share their findings, after which she introduces approaches to combating climate change, including mitigation (reducing and stabilizing the levels of heat-trapping greenhouse gases in the atmosphere), Adaptation (adapting to the climate change already in the pipeline), and Resilience (bouncing back from consequences of climate change, such as storms, droughts, etc.). She then asks students to revisit the UN website and search for others to identify solutions related to each sector. The teams research the following questions: 1) What are some solutions to climate change that can be addressed through each sector? 2) What data supports these claims? 3) Which solutions would be considered mitigation, adaptation, or resilience?

For this social studies lesson, Mr. Wiggins begins by musing, "If you are passionate about something and want to change it, how would you go about doing that?" He thinks out loud and shares a few examples: "I can write letters. I can certainly say something. I can join a group. What else can I do to advocate for change?" Student responses range from protesting, creating a banner or sign, filming a video, to creating a blog or a vlog. Reinforcing Ms. Vasquez's focus on climate change, Mr. Wiggins then asks his students to create their own representation of change. Students are free to choose any medium in which to create their own climate change advocacy statement. They can write a letter or a poem, create a placard or cartoon, compose a rap or song, draw or paint an image, or create a short video or video-based representation using their laptop and/or phone. With ten minutes remaining, each student is asked to briefly state (or show) their representation of change. Mr. Wiggins closes the lesson by stating,

"Remember, the First Amendment says you have the right to want to change things. You have the right to stand up and say, 'This is not what I believe in.' If climate change is something you believe in and are willing to stand up for, then you have the right to do so."

On the fourth day of the lesson (Elaborate), Ms. Vasquez informs the class that later in the day during social studies, they are going to participate in a climate summit simulation activity where they are going to work in groups to support one approach to addressing climate change. In science class, they are going to prepare for the summit. Ms. Vasquez then has students randomly count off 1 to 5, with each number being assigned to one of the five sectors/solutions. She then informs the groups that they are now delegates representing their position at a UN Climate Summit. Each team is then challenged to develop the following types of arguments for why this issue should be prioritized over the next five years:

- Two scientific arguments (what is or what will be): Provide data/evidence
- Two ethical arguments (what should be and why)
- Two solutions to address the issue (one mitigation and one adaptation/resilience)

Later in the day, at the UN Climate Summit, Mr. Wiggins explains that each group will be given a one-minute opening statement in which to present their key arguments. Groups will then have time to question, refute, or bolster their or others' arguments in a respectful manner. If time permits, he shares that they will try to negotiate an agreement and reach consensus. Mr. Wiggins uses a scoring rubric (see table 10.1), which he shares with the class in advance. The rubric evaluates each team's scientific and ethical arguments, understanding of mitigation/adaptation/resilience solutions, and their use of evidence and organization and presentation.

After the teams have a chance to articulate their arguments, ask questions, and negotiate an agreement through consensus (if time permits), Mr. Wiggins asks students, "What did you learn about the issue of climate change decision

Table 10.1 Scoring Rubric

Criteria/Score	0	1	2
2 Scientific Arguments			
2 Ethical Arguments			
2 Solutions (Mitigation and Adaptation/Resilience)			
Organization/Presentation			

making?" His students share that complex issues such as climate change mitigation and adaptation are not straightforward and have many possible "correct" solutions. Mr. Wiggins then asks, "Which arguments were most compelling to you? Why?" His students explain that arguments with strong scientific evidence and backed by reliable sources were very persuasive, but ethical arguments, particularly those that touched on highly emotional issues such as wildlife conservation or economics (e.g., "We have a duty to protect other species because they didn't cause climate change," or, "Imagine if your family worked in the oil business for generations and could lose their jobs and homes!") are also quite compelling. Mr. Wiggins explains this is why it is important to examine issues from a variety of perspectives and with a range of data sources to make informed decisions. Mr. Wiggins reminds his students of the CARS rubric[22] for assessing the quality of sources he had taught earlier in the year. CARS stands for credibility (e.g., What do we know about the author's credentials? Is the resource trustworthy?); accuracy (e.g., Is the resource current and comprehensive?); reasonableness (Does the resource seem fair and objective or highly biased?); and support (Does the resource cite its sources? Is the information corroborated across resources?). He then asks his students, "What additional evidence might you need/want to gain further understanding?"

On the fifth and final day of the lesson (Evaluate), Mr. Wiggins uses the scoring rubric to provide each team with an overall evaluation. As an extension that will be evaluated by both Ms. Vasquez and Mr. Wiggins, each student will produce a minimum three-hundred-word position statement that must include the following elements: a) an overview of at least two scientific arguments that resonated (or influenced) them the most; b) an overview of at least two ethical arguments that resonated (or influenced) them the most; and c) a succinct statement stating their proposed solutions, with one being mitigation and one being adaptation/resilience. Lastly, the position statement must provide at least two ways the student can (or will) take action to influence or implement the change(s) for which they advocate. This culminating project blends the inquiry-based empiricism of science with the participatory-based civic advocacy of social studies.

With so much interest in their lesson across grade levels, Ms. Vasquez and Mr. Wiggins share with their colleagues the following tips for adapting the lesson for a range of students.

Adaptations for Younger Students

- Start the lesson with a picture book to engage students, provide a common starting point, connect to English language arts standards, and spur discussion. Some book recommendations can be found in Chapter 4 of this book.

- Provide pictorial and textual cues for all activities, including the four corners discussion, the UN debate, and all data sheets.
- Read the UN Climate Change pages aloud as a class and discuss before having students complete their data sheets, or create grade-appropriate text versions of the webpages.
- Limit the number of websites students can use to develop arguments and provide some recommended sources (but you can still help students evaluate various sources using the CARS rubric).
- Provide detailed graphic organizers for the position paper or forgo the position paper and allow students to share their opinions through alternative means, including presentations, drawings/cartoons, plays, songs, or other forms of communication.

Adaptations for Older Students

- Allow students to identify additional sectors and solutions they might wish to study, beyond those provided on the UN website.
- Encourage students to work together to negotiate a final decision on climate change priorities and develop policies to address those priorities.
- Provide students with venues for sharing their position papers beyond your class, including school and local newspapers, class/school blogs, video posts, letters to state and local representatives, etc.

Best Practices Summary

Implementing SSI to address climate-change standards is a research-based approach that will reap many benefits for students, including increased content retention, real-world application of learning, enhanced perspective-taking skills, and hopefully, a sense of joy and wonder in both science and social studies. However, SSI requires a bit of preparation and facilitation on the teacher's part of the teacher. Here, we share some best practices to support those efforts.

- Provide students with a range of resources (e.g., websites, articles, videos) for their research and regularly model use of the CARS rubric to help students become experts in evaluating media sources.
- Remind students that climate change requires understanding and integrating both scientific and ethical arguments to develop comprehensive and effective approaches to mitigation and adaptation. There are no simple solutions!
- Encourage students to research and develop arguments that are contrary to their own initial positions in order to facilitate perspective-taking skills. Fol-

low this with an opportunity for students to express their own perspectives, grounded in evidence and bolstered by plans for action.
- Ensure students have the opportunity to "publicly" engage in evidence-based argumentation about climate change issues. Argumentation can take the form of debates, poster sessions, artwork, songs, public service announcements, class blogs/discussion boards, plays, and so on. Providing options for students to share their positions is both inclusive and empowering.
- Provide rubrics that clearly delineate your expectations for evidence-based arguments on climate change so assessment approaches and products such as those described in the prior bullet point are both equitable and rigorous.
- Encourage students to bring in articles, videos, and social media posts about climate change so the class can evaluate the quality of any data presented together and become familiar with the range of actions they can take as informed and engaged citizens.

Closing

Contemporary views of science and social studies education envision students as informed, engaged, and empowered citizens who can apply their knowledge to everyday life and societal decision making. Particularly regarding climate change, students' educational experiences must pique their curiosity, inspire them to grapple with complex issues, hone their perspective-taking skills, and propel them into action. We are confident the SSI framework's strong emphasis on evidence-based decision making, moral development, and application of content to real-world contexts make it particularly well suited to climate-change education and well positioned to develop the next generation of informed and empathetic thinkers, creators, leaders, and doers!

Notes

1. Douglas A. Roberts and Rodger W. Bybee, "Scientific Literacy, Science Literacy, and Science Education," in *Handbook of Research in Science Education, Vol. II*, ed. Norman G. Lederman and Sandra K. Abell (New York City: Routledge, 2011), 545–58.

2. "2020 New Jersey Student Learning Standards: Science," New Jersey Department of Education, May 10, 2024, https://www.nj.gov/education/standards/science/Index.shtml.

3. NGSS Lead States, *Next Generation Science Standards: For States, By States* (Washington, DC: The National Academies Press, 2013).

4. National Council for the Social Studies, *National Curriculum Standards for Social Studies: A Framework for Teaching, Learning, and Assessment* (Silver Spring, MD: NCSS, 2010).

5. National Council for the Social Studies, *The College, Career, and Civic Life (C3) Framework for Social Studies State Standards: Guidance for Enhancing the Rigor of K-12 Civics, Economics, Geography, and History* (Silver Spring, MD: NCSS, 2013).

6. "Climate Change Education by Grade Band," New Jersey Department of Education, January 2, 2024, https://www.nj.gov/education/standards/climate/learning/grade-band/.

7. Sami Kahn, *It's Still Debatable! Using Socioscientific Issues to Develop Scientific Literacy, K-5* (Arlington, VA: NSTA Press, 2019).

8. Dana L. Zeidler and Troy D. Sadler, "Exploring and Expanding the Frontiers of Socioscientific Issues," in *Handbook of Research on Science Education, Vol. III*, ed. Dana L. Zeidler et al. (New York City: Routledge, 2023): 899–929.

9. New Jersey Climate Change Education Hub, https://njclimateeducation.org/.

10. "Explainers: Transforming Climate Issues into Action," United Nations Climate Action, https://www.un.org/en/climatechange/science/climate-issues

11. "MS-ESS3-3 Earth and Human Activity," Next Generation Science Standards, https://www.nextgenscience.org/pe/ms-ess3-3-earth-and-human-activity

12. "MS-ESS3-4 Earth and Human Activity," NGSS, https://www.nextgenscience.org/pe/ms-ess3-4-earth-and-human-activity

13. "MS-ESS3-5 Earth and Human Activity," NGSS, https://www.nextgenscience.org/pe/ms-ess3-5-earth-and-human-activity

14. "MS-LS2-4 Ecosystems: Interactions, Energy, and Dynamics," NGSS, https://www.nextgenscience.org/pe/ms-ls2-4-ecosystems-interactions-energy-and-dynamics

15. "MS-LS2-5 Ecosystems: Interactions, Energy, and Dynamics," NGSS, https://www.nextgenscience.org/pe/ms-ls2-5-ecosystems-interactions-energy-and-dynamics

16. New Jersey Department of Education, "6.3 Active Citizenship in the 21st Century by the End of Grade 8," in *Draft 2020 New Jersey Student Learning Standards – Social Studies* (Trenton, NJ: 2020).

17. NJDOE, "6.3 Active Citizenship."

18. Rodger W. Bybee et al., "The BSCS 5E Instructional Model: Origins and Effectiveness" (Colorado Springs, CO: BSCS, June 12, 2006), 88-98.

19. Middle School Earth & Space Science – NGSS, "Earth's Changing Climate," *Khan Academy* video, 4:29, March 1, 2022, https://www.khanacademy.org/science/middle-school-earth-and-space-science/x87d03b443efbea0a:earth-and-society/x87d-03b443efbea0a:earths-changing-climate/v/earths-changing-climate

20. "Earth's Changing Climate," *Khan Academy*.

21. "Explainers," UN Climate Action.

22. Robert Harris, *WebQuester: A Guidebook to the Web* (Boston, MA: McGraw-Hill, 2000).

CHAPTER 11

The Role of Art in Climate Change Education

Carolyn McGrath

Humanity is at a crossroads. We must re-envision our relationships with each other and the more-than-human world or risk facing the worst impacts of anthropogenic global warming. The Intergovernmental Panel on Climate Change (IPCC) *Emissions Gap Report 2022* outlined the need for radical changes across all sectors of society: "Only an urgent system-wide transformation can avoid an accelerating climate disaster."[1] To meet the demands of this moment and to prepare young people for the future world they will inhabit, the education sector must also be part of this transformation.[2, 3] Comprehensive climate change education is critical if we are to halt and reverse the planetary emergency.[4]

If and when students learn about anthropogenic global warming, it is generally in science classes. Globally and nationally, these experiences need to be more consistent, accurate, and effective than they presently are, since robust science education is essential if we are to address the climate crisis.[5,6,7] At the same time, the multidimensional nature of the planetary emergency requires us to think outside traditional knowledge silos. We are living through a polycrisis of overlapping issues that include but are not limited to: species extinction and biodiversity loss; land, air, and water pollution; disproportionate environmental impacts on frontline communities; income inequality; pandemics; war and mass migration; and conflicts over natural resources. For young people especially, these issues also intersect with a crisis in mental health. If climate education is to be successful, it must address the full spectrum of ecological and social challenges and extend beyond the sciences to include the humanities, social sciences, and the arts. Cross-, inter-, multi-, and transdisciplinary pedagogies are urgently needed.

The visual arts have a critical role to play as part of an integrated approach to climate change education. Art allows young people to visualize and communicate climate change's causes, consequences, and potential responses. It is a powerful vehicle for storytelling that helps students make meaning through exploring

Figure 11.1 Climate Change Collage by Anastasia Angarone, Grade 9

and expressing their feelings about living on a warming planet. Art facilitates the development of empathy and connection to other people and the more-than-human world. It encourages students to develop a deeper understanding of the communities experiencing the worst consequences of climate change and demands that greater justice be served. Art creates outlets for advocacy and chances to have real-world impacts.

The process of art education also nurtures habits of mind that extend beyond the arts and are transferable to other fields of knowledge and practice. Art teaches creative problem-solving as well as inventive and critical thinking. It urges students to question, disrupt, and reimagine the status quo. Visual art is particularly good at cultivating skills of careful, close observation of and connection to the physical environment. It offers the potential for dialogue and collaboration. Art fosters the development of global awareness, civic responsibility, and citizenship. It invites people to consider other points of view and has the power to change hearts and minds.

Importantly, climate education through the visual arts provides opportunities not available through other disciplines. Art communicates through line, color, shape, and other visual means; it speaks through metaphor; it is emotive,

THE ROLE OF ART IN CLIMATE CHANGE EDUCATION 133

nonlinear, and often nonrational. Art gives physical form to otherwise abstract concepts, and it does so in a way that can circumvent resistance to more data-based types of communication.[8] Art opens pathways to engagement through our bodies and emotions. All of these make the visual arts a powerful ally in the quest to engage, educate, and activate young people.

I have been an art teacher in New Jersey public schools for over twenty-five years. During this time, I have also advised student environmental groups and been active in local and global fights for justice. Until recently, my work outside the art classroom and my work within it remained fairly separate. However, in the last several years, I started to question that distinction. With the release of the IPCC's 2018 *Special Report on Global Warming of 1.5 Degrees Celsius*[9] and subsequent global youth uprisings, scientists and young people protesting in the streets made it clear: We had a little over a decade to keep the planet's warming below 1.5 degrees Celsius. At the same time, Indigenous resistance at Standing Rock and the Movement for Black Lives reminded us the fight was not just about curbing emissions, but about much longer histories of colonization, institutionalized violence, and environmental injustice. It became increasingly clear that my teaching practices needed to evolve to include the relevance of these issues for my students and their art.

In the years since, bolstered by the 2020 New Jersey Student Learning Standards that support climate change learning in the visual arts, I began to develop and integrate lessons on climate change, biodiversity, land, water, and air pollution, and environmental justice into the art classes I teach. The lessons focused on: making art that helps students better understand climate change, its impacts, and potential solutions; creating meaning by exploring and expressing

Figure 11.2 Golden-Mantled Tree Kangaroo by Oishee Sinharay, Grade 9

their feelings; developing a deeper connection to the more-than-human world; cultivating critical consciousness around issues of environmental injustice; and experiencing artmaking as a form of climate action. I am hopeful that my insights will be useful to other educators, arts administrators, students, artists, scientists, community organizers, activists, and anyone else concerned with how creativity can be a formidable force in the fight for a just and livable future.

In this chapter, I will explore the evolving role of art in environmental and climate change education in a rapidly warming world. Then, I will share some of the lessons I have developed and integrated into my art classes at the high school level, including student reflections on their experiences exploring different aspects of the climate crisis through art.[10]

Art and Art Education in a Changing Climate

While a specific focus on climate change is a more recent development in arts education, the broader field of eco- or environmental art education has existed for at least half a century, mirroring the concurrent rise of ecological consciousness and environmental art in the 1960s and 1970s. This period, marked by significant environmental milestones such as the first Earth Day and the establishment of the Environmental Protection Agency, also saw the emergence of land and environmental artists like Robert Smithson, Ana Mendieta, and Alan Sonfist. Eco-art education co-evolved alongside these societal and aesthetic movements, making way for a reimagined role of art education in environmental advocacy. According to art education scholar Hilary Inwood, "Eco-art education integrates knowledge, skills, values and pedagogy from the visual arts, art education and environmental education as a means of developing awareness of and engagement with environmental concepts and issues such as place, interdependence, systems-thinking, biodiversity, and conservation."[11]

In the years since the development of eco-art education, public awareness of climate change, biodiversity loss, pollution, and environmental justice has increased, while environmental artists have continued to address these concerns through their work. In 1985 and 1988, climate scientist James Hansen raised the alarm in testimony before Congress about the threat of global warming. The IPCC released its first scientific assessment of climate change in 1990. That same year, Robert Bullard wrote *Dumping in Dixie: Race, Class, and Environmental Quality*. In 1991, artist Mel Chin created *Revival Field*, a sculpture on a Superfund site and landfill, using plants as hyperaccumulators to remedy contaminations in the soil. Artist and writer Suzi Gablik released *The Re-Enchantment of Art*, a book advocating for art that is both socially responsible and ecologically engaged. In 1992, the United Nations Earth Summit took place in Rio de

THE ROLE OF ART IN CLIMATE CHANGE EDUCATION 135

Janeiro, establishing the United Nations Framework Convention on Climate Change (UNFCCC). At the summit, environmental artist Agnes Denes' site-specific work, *A Living Time Capsule-11,000 Trees, 11,000 People, 400 Years*, was presented as "Finland's contribution to help alleviate the world's ecological stress."[12]

The 1990s also saw a surge of interest in eco-art education. Art education scholars Douglas Blandy and Elizabeth Hoffman argued in 1993 for an "'art education of place' with which art educators can imagine new relations among art, community, and environment."[13] In 1996, art education professor Don Krug organized the Art & Ecology Colloquium at Ohio State University to question "pre-existing premises about linkages between art and ecology and suggested implications for art education curriculum."[14] The next year, Krug served as guest editor of the National Art Education Association's *Art Education* journal, which also focused on the theme of art and ecology.[15] Doug Blandy, Kristin Congdon, and Don Krug wrote about the need for "contemporary examples that art educators and their students can look to as they formulate ecologically restorative projects"[16] and held up the radically interdisciplinary artmaking of Helen Mayer Harrison and Newton Harrison as an exemplary practice.

By the mid-2000s, the realities of the climate crisis were becoming increasingly apparent. In 2005, Hurricane Katrina hit the Gulf Coast, devastating poor Black communities that were already struggling with the legacies of environmental racism.[17] In 2005, writer Bill McKibben (who established the climate activist organization 350.org in 2008), famously wrote in *Grist Magazine*: "What the Warming World Needs Now is Art, Sweet Art."[18] In 2006, Al Gore's groundbreaking film *An Inconvenient Truth* starkly warned about the dangers of continuing to burn fossil fuels. In 2007, Eve Mosher began her site-specific work

Figure 11.3 Nyombi Morris, Climate Hero by Eleanor Murray, Grade 12

High Water Line, marking areas in Manhattan that would flood just a few years later during Superstorm Sandy.

That same year, the National Art Education Association's *Studies in Art Education* journal featured a special issue on eco-responsibility in art education, including multiple investigations in place-based environmental art education. Laurie Hicks and Roger King wrote in the introduction, "Artists and educators can and must play a role in bringing about a more environmentally responsible and ecologically literate culture.[19]" At the end of the decade, graffiti artist Banksy ominously spray painted the words "I don't believe in global warming," partially submerged in a London canal, in an apparent response to failed climate talks in Copenhagen in 2009.

Since the aughts, we have experienced significant and profound changes to the earth's atmosphere. In 2010, carbon dioxide was 390 parts per million (ppm). In 2023, it was 424 ppm, a concentration that has not existed for millions of years.[20] We are on track to cross the 1.5-degree threshold established by the 2015 Paris Climate Accord by the 2030s.[21] The ten hottest years on record have all been since 2010, and 2023 was the warmest year ever documented.[22] The 2010s in the United States saw climate-fueled Hurricanes Sandy, Harvey, and Ida wallop the East and Gulf coasts as drought and fires raged in the West. Black, Brown, Indigenous, low-income, and other historically under-resourced communities continued to experience the worst impacts of a changing climate.

Figure 11.4 New Jersey State Bird: 1935: American Goldfinch, 2030: ? by Melanie Edwards, Grade 9

On a global scale, developing nations continued to pay the price for the high emissions of wealthy countries.

At the same time, the movement for climate justice was also strengthening, whether through Indigenous communities fighting for sovereignty in the face of expanding oil infrastructure or the Black Lives Matter movement drawing connections between racialized violence and the disproportionate environmental impacts faced by people of color. In 2014, the People's Climate March brought these and other groups together under the banner of "To Change Everything, It Takes Everyone" for the largest climate protest in history. By 2018, Greta Thunberg's Fridays for Future sparked an international youth movement whose school strikes and massive street protests demanded leaders take bold, swift, and necessary climate action. During these years, influential works of climate and environmental justice art were also gaining notice. In 2015, artist Olafur Eliasson and scientist Minik Rosing arranged melting Greenland glaciers in the form of a clock outside the Place du Panthéon for the Conference of the Parties (COP21) in Paris. Ruby LaToya Frasier documented the Flint water crisis in her 2016 photo series *Flint Is Family*. John Acomfrah explored the legacies of pollution, extractivism, and colonial conquest in his poetic multi-screen film *Purple* in 2017. Maya Lin continued to produce art as part of *What's Missing*, an ongoing multi-site memorial on species extinction, climate change, and climate solutions.

Art education scholars have continued to insist on the central importance of art in climate change and environmental education. In 2012, art educators Tom Anderson and Anniina Suominen Guyas envisioned "a pedagogical approach built on the cognition that to stop injustice, cruelty, and discrimination and to work toward a more humane society, we need to [...] acknowledge the interconnectedness social justice and environmental issues."[23] In the 2012 *Studies in Art Education* special issue entitled "Sustainability and What it Means to Be Human," Charles Garoian asked his reader to "imagine if the object of art education in the 21st century was to create ecological alliances, to serve as a node" of true cross-curricular study.[24] Art education scholar Joy Bertling has written for several years about the value of critical place-based environmental art education and, more recently in *Art Education for a Sustainable Planet*, outlined approaches that combine eco-art and eco-pedagogical approaches. Interdisciplinary social scientist Julia Bentz has been championing the transformative potential of teaching climate change through art.[25]

For more than fifty years, there has been a clear and consistent call to action by artists as well as art education and environmental scholars to address the social and ecological crises of our time through environmental art education. Unfortunately, despite these efforts, impacts in actual art classrooms have been limited. In *Mapping Eco-Art Education*, Hillary Inwood wrote in 2008, "While artists have been devising creative solutions to environmental problems since the 1970s, art educators for the most part have not kept pace, and have not done enough to share their work with a broader audience."[26] In a 2017 study, Jeniffer and Doreen Sams found US educators faced numerous barriers to integrating environmental education into art classes.[27] In 2021, Joy Bertling and Tara Moore investigated the extent to which art educators incorporated ecological themes into their lessons and found that while many wanted to, they often prioritized other areas or approaches instead.[28] Lack of teacher training and professional development was a consistent theme across studies.

The literature describing the lag in integrating environmental, climate, and justice concerns in art education mirrors my own experiences as an art teacher. Although I have been very aware of and engaged with ecological and social issues outside of the classroom, I only started actively centering these topics in my art lessons in the last several years. Certainly, some of this reflects the type of training I received in my art education program, which favored Discipline-Based Arts Education (DBAE) and multicultural arts education approaches. While neither of these frameworks excluded the incorporation of environmental issues, I did not consider the possibility that I could teach art as a means to a more sustained investigation of ecological and related social concerns. Indeed, although I was quite knowledgeable about environmental problems and environmental art, I was generally unaware of the long history of scholarship advancing environmental education *through* art. Other barriers included the unspoken rules that define what is considered discipline-specific knowledge, especially at the highschool level. I did not want colleagues to question me being out of my departmental zone.

While educators like myself (both in and outside the sciences) have been hesitant to integrate urgent climate, environmental, and justice issues into our lessons, the global problems we have known about for decades continue to worsen and accelerate. It is incumbent upon all educators and educational institutions to find ways to meaningfully integrate these concerns into our classrooms swiftly and meaningfully. I have, to the best of my ability, spent the last several years doing so.

Ways of Teaching Climate and Environmental Education Through Art

As indicated in my summary, the field of environmental and climate art education is broad and includes diverse approaches and emphases. While my overview was not exhaustive, it hopefully gives a sense of some of the strands and directions that have developed over the years. It is beyond the scope of this chapter to delineate the distinctions between all the different approaches, but I would

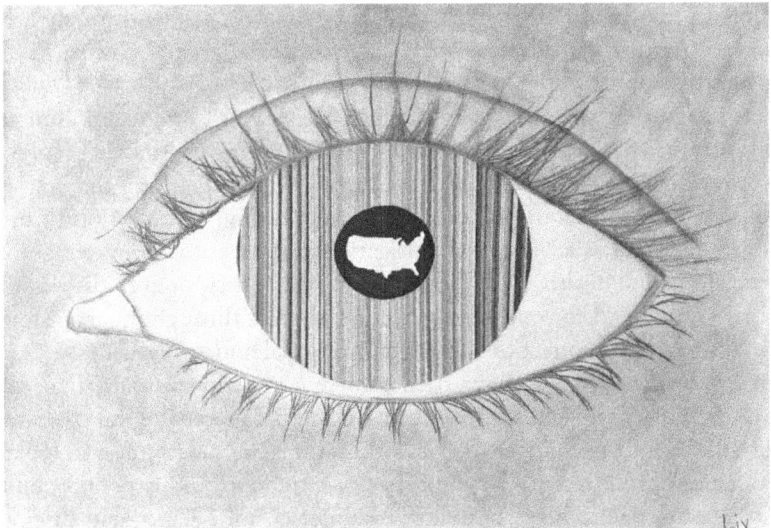

Figure 11.5 Global Warming Stripes, United States by Olivia Loutfi, Grade 10

like to outline some trends in environmental and climate art education I believe should be approached with caution and others that I fully embrace.

When many art teachers consider integrating climate and/or environmental issues into their classes, they choose to have students create art from waste, such as making a mural from bottle caps. While this can teach the vital ecological concept of "reuse" and help students feel good about preventing items from going to a landfill, upcycling materials does not inherently help students learn about larger systems of waste and disposability. It does not draw attention to who is responsible, who profits, and who suffers as a result, and it does not identify the broader -climate, ecological, and social impacts of those systems. Reusing items in an art room might be a way to begin exploring these issues, but more often than not, it fails to challenge the status quo. At worst, it can make students feel it is their job to find uses for all the waste created in the world, which is a burden no young person should have to bear.

Other approaches that risk remaining superficial include making art from natural materials. There is certainly great value in introducing students to land artists such as Andy Goldsworthy or Nils Udo and their methods for working with materials from the surrounding environment. However, without deeper explorations into the relationship of humans to the landscape (and the degradation of it) or the inseparability of people from the more-than-human world, these investigations are in danger of romanticizing nature, reinforcing our separation from it, and sidestepping the damage we've inflicted upon it and each other.

Finally, another approach to teaching climate change through art that should be carefully considered is making art that is "about" nature without a larger interdisciplinary critical component. For example, my Art Fundamentals students are currently drawing and painting tree leaves from around our school campus. On its own, it is a valuable study in observational drawing, color mixing, watercolor painting techniques, and using objects from the immediate environment as subjects of one's art. However, without exploration of broader issues around trees, such as their mitigation effects on stormwater and heat; their social, mental health, and biodiversity benefits; their uneven distribution in wealthier vs. redlined communities; or even their role throughout art history, the lesson fails to draw connections to larger ecological and social issues.

In my own teaching, I have tried to take a holistic approach that emphasizes the web of relationships involved in any climate or ecological concern. We cannot talk about the changes to the earth's atmosphere without also looking at related issues of biodiversity loss and species extinction, industry and pollution, and social injustice. This type of cross-, inter-, multi-, and transdisciplinary perspective offers numerous opportunities and benefits for both educators and students, many of which I have outlined at the start of this chapter.

While my approach is holistic, I also have structured my climate and environmental lessons around specific intentions. Loosely, these include one or more of the following frameworks:

1. Climate literacy: Art provides a means of deepening climate change and ecological knowledge.
2. Meaning-making: Art creates an opportunity to explore feelings about climate and other ecological issues through storytelling.
3. Biophilic: Art facilitates the development of love, empathy, and connection toward the human and more-than-human world.
4. Climate justice: Art allows the development of critical consciousness around issues of injustice and becomes a means of resistance.
5. Climate action: Art is a form of climate action that provides experiences of empowerment and efficacy.

General Recommendations for Introducing Climate Change in Art

Before integrating climate change and related environmental/social issues into an art lesson, it is helpful to first assess students' current level of understanding about these issues. If, like me, you teach a variety of ages within the same class, students probably have had different experiences of climate education in the past. Some might be quite knowledgeable, while others may have had little exposure. It is also instructive to inquire what your students are curious to know more about. What questions do they have? Knowing where you are starting from and where your students want to go can help you better design and implement your lessons.

I have developed a short introduction to the causes, impacts, and responses to the climate and biodiversity crisis that I present at the start of lessons on these topics. I tailor it to the needs of each individual class, based on what they already know and what they want to know more about. During this introduction, I make sure not to simply focus on the challenges but also emphasize the numerous ways people around the world are working to confront these problems. We discuss concepts like climate and environmental justice and talk about how people of the global majority are fighting to right these wrongs. I share numerous examples of collective action and the ways young people in particular are making a difference.

This is a good place to introduce artists who are addressing social and ecological challenges in their work and who use artmaking as a form of climate action. Students may be unfamiliar with the history of environmental art or even the concept that art can be a means to explore contemporary issues. In my classes, depending on the focus of the lesson, I have shared the work of many of the artists I have referred to throughout this chapter, such as Agnes Denes, Mel Chin, Eve Moser, John Akomfrah, Olafur Eliasson, and Maya Lin. I have also presented the work of other artists who explore climate in their art, such as Zaria Forman, Jill Pelto, Nicholas Galanin, Daniela Molnar, Allison Janae Hamilton, Courtney Mattison, Mary Mattingly, Aïda Muluneh, and others. The beginning of a climate art lesson is also the perfect time to highlight other arts, such as the poetry/spoken word of Amanda Gorman's "Earthrise," Childish Gambino's song "Feels Like Summer," or dance performances by the Artichoke Dance Company. Students may be especially interested to learn about local art and artists who are responding to climate issues, such as the Mural Arts Philadelphia Climate Justice Initiative,[29] the site-specific climate art created at the New Jersey shore,[30] or Ecological City, the multidisciplinary art and climate solutions celebration that takes place annually in Manhattan.[31] It might surprise students to learn there are even museums and art shows dedicated to climate change and its solutions.[32, 33]

The Lessons

Over the last several years that I have been teaching climate through art, I have introduced a variety of different lessons in different classes with different student groups. In this section, I will give an overview of several of those lessons, what they entail, and their outcomes.

ENVIRONMENTAL COLLAGE

The environmental collage lesson took place in my Art Fundamentals class. After a short presentation on climate change, biodiversity loss, ocean plastic pollution, and environmental justice, students had the opportunity to select one area they wanted to research further. I provided them with links to curated sources where they could learn more about the topic they chose. We then explored the history of collage and how artists like Hannah Höch, Martha Rosler, and Wangechi Mutu work/ed with this art form to address the social and political problems of their time. Students drew on their research to guide the creation of their own collage, using cut magazines and techniques like juxtaposition to highlight their points of view. Many students highlighted threats to mammals and aquatic life, with animals embedded in swathes of plastic and nets, struggling, or even dead. Some students also included human subjects and emotions. One student showed tears falling on a landscape of ocean debris, and another showed eyes filled with fire, spilling out onto melting glaciers. Several students chose media such as cameras and televisions, emphasizing the ways we see and consume many types of environmental degradation. Images of money appeared in some students' collages, highlighting the connection between ecological destruction and profit. Finished collages were shared with the class and formed the basis of a more in-depth discussion on art, justice, and environmental issues.

ENDANGERED SPECIES ART

The endangered species lesson occurred in my World Arts class during a unit on art from India. We began by exploring Madhubani art, as well as other forms of traditional painting from India that emphasize the interrelationship between people, animals, and the environment. We learned about how Madhubani artists even painted on trees to prevent them from being cut down.[34] I talked about how Indigenous people globally are protectors of biodiversity,[35] and we discussed how artists like Maya Lin have used their art as a kind of environmental intervention. Students then chose an animal that is vulnerable to threat or extinction, conducted independent research, and developed a painting inspired by colors and patterns in traditional Madhubani art. The artworks depicted a variety of threatened or endangered animals, such as the golden-mantled tree kangaroo, poison dart frog, blue whale, whooping crane, and the Bornean orangutan. Paintings were highly stylized and used bright, high-contrast colors and repeated patterns. Once complete, the students displayed their art in our town's public library, along with cards explaining the animals, their characteristics, and threats. Library patrons could write feedback to the student artists in a guest book after viewing the exhibition. Students also presented their artwork and research to each other in the form of a class presentation and discussion.

CLIMATE HERO PORTRAIT

The climate hero portrait lesson took place in my Art Fundamentals class. The lesson began with a short overview of climate change causes, consequences, and responses. Then, I shared several artists who address climate change through their work. (See General Recommendations for Introducing Climate Change in Art.) Next, I talked with students about the young people who have been at the forefront of the movement for climate justice. Most students have heard of Greta Thunberg, but they may not be aware of the many other powerful youth climate activists from around the world, such as Vanessa Nakate, Autumn Peltier, Xiuhtezcatl Martinez, or Xiye Bastida. Students then explored a Padlet I created with introductions to these and many other youth leaders. Next, we discussed artists like Maliha Abidi and Shepard Fairey who made portraits of young climate activists to raise awareness of their messages. Finally, students chose a youth climate leader whose work they wanted to amplify through the creation of a portrait with symbolic elements. The artworks resulting from this lesson were varied, as students chose their own art media. Some portraits were created with drawing or colored pencils, others were done with watercolor or acrylic paint, and some were made with mixed media. After this lesson was complete, I shared the artwork on social media and tagged the youth climate leaders whose portraits the students had created. Several of the activists responded, expressed deep appreciation for the art, and reshared the work through their own social media networks. Nyombi Morris, a youth climate leader from Uganda, even ended up visiting our school and talking about his climate activism after one student shared a portrait she had created of him.

NEW JERSEY: PROTECT WHAT YOU LOVE

New Jersey: Protect What You Love is a ceramic tile lesson I taught in my Ceramics class. I introduced the project with a brief presentation on climate science and the consequences of climate change both globally and in New Jersey. We then discussed solutions that have been proposed or are currently being implemented. I asked students to reflect on the people, places, and experiences in New Jersey that were meaningful to them and to share the impact climate change has had or will have on the things they love. Students used the climate change section of the New Jersey Department of Environmental Protection website, along with other suggested resources to further investigate how climate change is impacting the state. By creating a tile that communicates aspects of what they love in New Jersey, students could educate and inspire others to take action. Many of the art pieces centered on the birds, animals, and plants that would be affected. Several highlighted the impacts of climate change on the New Jersey shore, a place to which many students have personal connections. Students photographed their tiles, and I shared images of them through social media. The class also had the opportunity to see and discuss each other's work, as well as the issues they brought up.

ART X CLIMATE

While most of the lessons I shared have taken place in my classes, I also have explored climate themes in the art club I advise. Last year, there was an opportunity for students to submit artwork for the Fifth National Climate Assessment, the first time this US government report included art from both adult and youth artists. Eager to have students' work considered, but knowing I did not have time to devote to it in my art classes, I proposed students in Art Club take it on as a project. I approached the lesson much like in my classes, sharing a basic introduction to climate change causes, impacts, and responses. We then investigated the different parts of the report, and students identified something personally relevant to them. Over the next several weeks, I helped students develop and refine their ideas and set them up with the art materials they needed. Students' finished art engaged with a variety of themes, although the mood of many pieces was somber. One student artist depicted a young person creating a colorful work of art while factories and grey smoke-filled skies loomed in the background. A different student painted a landscape with cows grazing in a meadow, the sky hanging heavy with ominous green-grey clouds. A third artwork showed a northern right whale untangling itself from a net. Another student used the global warming stripes and a map of the United States to fill the interior of a large eye. Remarkably, three of these student pieces were selected for inclusion in the *White House Climate Report*, alongside the work of recognized climate artists such as Xavier Cortada and Jill Pelto. Another was chosen for publication in an environmental magazine.

Discussion and Student Reflections

Considering the five frameworks I outlined earlier, all the projects fulfilled multiple overlapping expectations. They all deepened climate and biodiversity knowledge (climate literacy framework). They all facilitated students' exploration of feelings about climate and ecological issues through storytelling. Several lessons addressed students' love for place, people, and the creatures that inhabit the more-than-human world, while the endangered animal project considered this most directly. Climate justice was woven throughout all the lessons, but students had the most opportunity to consider it through the environmental collage, the climate hero portrait, and New Jersey: Protect What You Love. While the climate hero portrait project most explicitly used a climate action framework, all the lessons provided outlets for climate action through publicly sharing completed artwork (if only in a class setting and/or through social media).

As indicated previously, I make a practice of soliciting student feedback before these lessons. I also asked students for reflections during the assignments and after they had completed their artwork. Their responses are worth further consideration through the lens of the five frameworks (climate literacy, meaning-making, biophilic, climate justice, and climate action).

Before the lessons, most students seemed to have a general understanding of the basics of climate change, although some were unclear or had misconceptions about specifics. Most students (97 percent) understood climate change to be real, but only about 45 percent knew there is a scientific consensus. About 70 percent of students were uncertain whether people in the United States care about climate change. Around 90 percent knew there are things that can be done to lessen the impacts of climate change. Approximately half of the students said most of what they knew about climate was learned through social media. After the lesson introductions, I questioned students again. Nearly 80 percent agreed or strongly agreed that they now knew more about the causes, impacts, and solutions to climate change; 60 percent felt empowered to take action; and 80 percent said they understood the role art could play in addressing climate change.

Several students remarked in their narrative reflections on specific things they had learned about climate change, climate justice, and/or biodiversity through making their artwork. One noted that she had not previously considered the extent to which climate change was already impacting New Jersey: "Beaches getting shut down because of erosion, extreme weather, animals migrating to cooler climates, and the air getting polluted are happening all around our state." A different student reflected, "I knew that climate change was real, it was happening, and I could be doing something about it. But I never realized that it's affecting me, my neighbors, all of New Jersey, and everywhere in the whole world right now."

For many students, making the artwork was an opportunity to process, make meaning of, and express their feelings. Their emotional response to the changes happening around them showed many parallels to the findings of the 2021 Lancet global study of ten thousand young people's emotions about climate change.[36] The Lancet research indicated that most youth around the world feel sad, anxious, angry, powerless, helpless, and guilty. Many young people also feel a sense of betrayal due to government inaction. My students reflected:

- "I feel worried, sad, and anxious about my future."
- "I have been sad about the careless treatment of the environment for my whole life, and I will continue. And I'm afraid of what's coming next for the world going forward."
- "The pressing nature [of climate change] adds stress to my life since it seems that the governments/agencies in charge of fixing it aren't doing it fast enough."
- "Sometimes I feel that combating climate change is impossible and that we have no hope."
- "It's frustrating to see people in positions of power not doing anything to try and solve the climate crisis."
- "Most of the time, climate change is something that leaves me scared about the future of the planet, as well as my generation and humanity itself."
- "My feelings regarding the climate crisis are a sense of fear paired with a sense of hope."

While the range of feelings evoked by the ecological and climate emergency weighed very heavily on students, having the chance to make meaning of those emotions through art was an immensely powerful and positive experience.

- "Learning about [my climate hero] and making this portrait made me feel proud."
- "While I made my art, I felt happy to focus on a subject matter that I love."
- "Learning more about [my climate hero] led me to feel inspired."
- "Making a portrait of someone not much older than me who is making a profound impact on the climate crisis was quite relieving."
- "This project did have a very big effect on my perspective of climate change, and it makes me more determined to speak up about it."
- "It made me feel very proud of my work and glad that my message was out there."
- "I am glad that we did the project because I learned from it and felt as though I was spreading awareness about something I was passionate about."

Students also appreciated the opportunity to deepen their connection to the human and more-than-human world through their artmaking.

- "I wanted to communicate the joy and innocence of the crayfish and how they don't deserve to lose their homes."

THE ROLE OF ART IN CLIMATE CHANGE EDUCATION   147

- "I chose to do my tile about the goldfinch because it is the favorite of many people all throughout the state, including my mom and I."
- "I chose the orangutan because I feel such a close connection to that animal. I also feel like that connection is due to the fact that orangutans and apes in general resemble so many parts of us."
- "[Making the painting] made me feel so connected to the North Atlantic right whale. It just hurt so much knowing that their numbers could continue to decline."
- "There are people who are genuinely struggling due to climate issues. I feel for the people who are experiencing these difficult climate conditions."
- "Climate change is not just affecting the environment and the atmosphere, but it affects human beings in a very real, personal way."
- "Creating the portrait allowed me to think about how climate change has affected the people in my community and those all around the world."
- "Hearing someone tell their story about how the climate crisis is affecting real people, families, towns, etc., reminds you that it's not just a news story, it's real life."

As students learned more about the disproportionate impacts of climate change, they began to reflect more deeply on this issue, develop greater empathy for those impacted, and think more critically about structures that cause injustice.

- "[My climate hero] lived near a waste treatment plant for the majority of her childhood and part of her adult life, and that caused really bad health problems for her that still affect her even after moving to a less polluted area. People who live in a place like we do don't see the effects of climate change in their everyday lives in the same way."
- "Countries such as Pakistan are struggling immensely, especially because they don't have the resources they need. Pakistan lacks clean water and has multiple power outages per day. If the climate crisis isn't resolved soon, people whose basic needs aren't met will begin to die or suffer. And that isn't okay."
- "[My climate hero] is from an area that is very undeveloped, and they rely on the land to live. This is why climate change has impacted her and her people a lot more than it has affected me. After researching [my climate hero], I realized . . . how it affects other people."

The last impact of these lessons, and perhaps the most important one, is the climate action framework. Most, if not all, of the students expressed that making art about the climate crisis gave them a sense of agency and a feeling of contributing to positive change.

- "[The art project] made me believe that if there is something I am really passionate about, I should stand up for it. It doesn't matter what age I am, even if it's ten years from now, using your voice to speak your opinion is vital."
- "I think that making this art did make a difference. It's not just about convincing a viewer to take action. It's about connecting with the heart of the viewer."
- "These climate change tiles can help share the word about the things you care about the most, which is one of the best things you can do in order to help spread awareness."
- "It has encouraged me to continue using my passion for art to speak out about what I believe in."
- "My peers and I can make a difference, similar to our climate heroes."
- "Having our artwork and their meanings displayed for the public to see felt like it made the greatest difference to me. Everyone who saw our paintings got the message that we were spreading, and if our project helped to inspire even one person to make their own stand, then I think we made a difference."
- "I have always wanted to use my art for the greater good, and working on this project made me feel that my actions were helping something useful and important."
- "I'm so glad I was given the opportunity to show my art to the world. It makes me feel very happy."

Several students remarked that creating the artwork encouraged them to make changes in their own lives.

- "I felt inspired to take more action in my own life regarding climate change."
- "It inspired me to involve myself more in conservation and education around the community."
- "Making the project pushed me more toward climate activism."
- "I have become more conscious about how I am helping the crisis and have taken more sustainable measures in my life."
- "I learned about the action people were taking publicly. I guess it hadn't occurred to me that I could go to protests, send letters to politicians, etc. I think these newly discovered forms of justice and action would definitely make my personal impact a little bit bigger."

Conclusion

The ecological crisis demands a radical rethinking of education. As we face a planetary emergency that impacts all aspects of our lives, teachers must transcend the limitations of disciplinary boundaries and embrace cross-, inter-, multi-, and transdisciplinary approaches. Artmaking can play a critical role in this process, as genre-defying environmental artists have been demonstrating for decades. Innovative art education scholars have long urged teachers to follow suit by inviting students to grapple with the urgent ecological and social issues of our time through art. Artmaking provides a powerful means for students to gain a deeper understanding of the polycrisis: It offers an opportunity to create meaning while exploring and expressing emotions; it facilitates the deepening of bonds to the human and more-than-human worlds; it challenges students to see and resist systems of injustice; and it is a powerful pathway for climate action. Through my own integration of lessons on climate change, biodiversity, pollution, and environmental justice in art, I have attempted to find ways to facilitate meaningful student engagement on these issues. My hope is that these reflections will be of use to others who choose to walk along this path and will inspire others to join.

Notes

1. United Nations Environment Programme, *Emissions Gap Report 2022* (Nairobi, Kenya: UNEP, October 27, 2022), https://www.unep.org/resources/emissions-gap-report-2022.
2. Christina Kwauk, "Is Education Standing Up to the Task of Climate Action?" Brookings Institution, September 18, 2019, https://www.brookings.edu/articles/is-education-standing-up-to-the-task-of-climate-action/
3. United Nations General Assembly, Resolution 70/1, *Transforming Our World: The 2030 Agenda for Sustainable Development* (September 25, 2015), https://sdgs.un.org/2030agenda
4. UNESCO, *Education for Sustainable Development: A Roadmap* (Paris: UNESCO, 2020), https://unesdoc.unesco.org/ark:/48223/pf0000374802.locale=en
5. Katie Worth, *Miseducation: How Climate Change is Taught in America* (New York City: Columbia Global Reports, 2021), https://globalreports.columbia.edu/books/miseducation/
6. Christina Kwauk, "Who's making the Grade on Climate Change Education Ambition?" Brookings Institution, November 5, 2021, https://www.brookings.edu/articles/whos-making-the-grade-on-climate-change-education-ambition/
7. National Center for Science Education and Texas Freedom Network Education Fund, *Making the Grade? How State Public School Science Standards Address Climate Change* (Oakland, CA & Austin, TX: NCSE & TFNEF, October 2020), https://ncse.ngo/files/MakingTheGrade_Final_10.8.2020.pdf
8. Nan Li et al., "Artistic representations of Data Can Help Bridge the US Political Divide Over Climate Change," *Communications Earth & Environment* 4, no. 195 (2023):1–9, doi:10.1038/s43247-023-00856-9. See especially Fig. 2: Mockup Instagram

Posts Used to Gauge Perceived Credibility, Perceived Relevance of Climate Change and Information Recall, https://www.nature.com/articles/s43247-023-00856-9#Fig 2

9. Intergovernmental Panel on Climate Change, *Special Report: Global Warming of 1.5°C* (Geneva: IPCC, 2018), https://www.ipcc.ch/sr15/

10. Note that I will henceforth primarily use the terms *climate, climate change, climate crisis* and related terms to refer to the range of intersecting environmental and social issues that define this moment in time.

11. Hilary J. Inwood, "Cultivating Artistic Approaches to Environmental Learning: Exploring Eco-Art Education in Elementary Classrooms," *International Electronic Journal of Environmental Education* 3, no. 2 (2013): 129-1–45, https://files.eric.ed.gov/fulltext/EJ1104868.pdf

12. http://www.agnesdenesstudio.com/works4.html

13. Douglas Blandy and Elizbaeth Hoffman, "Toward an Art Education of Place," *Studies in Art Education* 35, no. 1 (Autumn 1993): 22–33, doi:10.2307/1320835

14. Neperud, Ronald W. "Art, Ecology, and Art Education: Practices & Linkages." *Art Education* 50, no. 6 (1997): 14–20. https://doi.org/10.2307/3193683.

15. *Art Education* 50, no. 6, https://www.jstor.org/stable/i360128

16. Doug Blandy, Kristin G. Congdon, and Don H. Krug, "Art, Ecological Restoration, and Art Education," *Studies in Art Education* 39, no. 3 (Spring 1998): 230–43, doi:10.2307/1320366

17. Julie Sze, "Toxic Soup Redux: Why Environmental Racism and Environmental Justice Matter After Katrina," Social Science Research Council, June 11, 2006, https://items.ssrc.org/understanding-katrina/toxic-soup-redux-why-environmental-racism-and-environmental-justice-matter-after-katrina/

18. Bill McKibben, "What the Warming World Needs Now is Art, Sweet Art," *Grist*, April 22, 2005, https://grist.org/culture/mckibben-imagine/

19. Hicks, Laurie E., and Roger J. H. King. "Guest Editorial: Confronting Environmental Collapse: Visual Culture, Art Education, and Environmental Responsibility." *Studies in Art Education* 48, no. 4 (2007): 332–35. http://www.jstor.org/stable/25475839.

20. "Broken Record: Atmospheric Carbon Dioxide Levels Jump Again," National Oceanic & Atmospheric Administration, June 5, 2023, https://www.climate.gov/news-features/feed/broken-record-atmospheric-carbon-dioxide-levels-jump-again

21. Nicola Jones, "When Will Global Warming Actually Hit the Landmark 1.5° Limit?," *Nature* 618 (2023): 20, doi:10.1038/d41586-023-01702.w

22. Raymond Zhong and Keith Collins, "See How 2023 Shattered Records to Become the Hottest Year," *New York Times*, January 12, 2024, https://www.nytimes.com/2024/01/09/climate/2023-warmest-year-record.html

23. Tom Anderson and Anniina Suominen Guyas, "Earth Education, Interbeing, and Deep Ecology," *Studies in Art Education* 53, no. 3 (Spring 2012): 223–45, https://www.jstor.org/stable/24467911.

24. Charles R. Garoian, "Sustaining Sustainability: The Pedagogical Drift of Art Research and Practice," *Studies in Art Education* 53, no. 4 (Summer 2012): 283–301, https://www.jstor.org/stable/24467918.

25. Julia Bentz, "Learning about Climate Change in, with and through Art," *Climatic Change* 162 (2020): 1595–1612, doi:10.1007/s10584-020-02804-4.

26. Hilary Inwood, "Mapping Eco-Art Education," *Canadian Review of Art Education* 35 (2008), https://files.eric.ed.gov/fulltext/EJ822675.pdf

27. Jeniffer Sams and Doreen Sams, "Arts Education as a Vehicle for Social Change," *Australian Journal of Environmental Education* 33, no. 2 (July 2017): 61–80, https://www.jstor.org/stable/26422961.

28. Joy C. Bertling and Tara C. Moore, "A Portrait of Environmental Integration in United States K-12 Art Education," *Environmental Education Research* 27, no. 3 (2020): 382–401, doi:10.1080/13504622.2020.1865880.

29. https://www.muralarts.org/artworks/climate-justice-initiative/

30. StateoftheArtsNJ, "Climate Art in Fourt Acts," YouTube video, 6:20, November 1, 2021, https://www.youtube.com/watch?v=LN6kgV4TVkc&t=8s&ab_channel=StateoftheArtsNJ

31. https://earthcelebrations.com/ecological-city-project/

32. https://climatemuseum.org/

33. Louis Bury, "On Getting Things a Little Less Wrong," *Hyperallergic,* November 8, 2023, https://hyperallergic.com/855363/on-getting-things-a-little-less-wrong-climate-futurism-pioneerworks/

34. Amarnath Tewary, "Indian Tribal Art Form Madhubani to Save Trees," *BBC News,* November 28, 2023, https://www.bbc.com/news/world-asia-india-20422540

35. Steve Nitah, "Indigenous Peoples Proven to Sustain Biodiversity and Address Climate Change: Now It's Time to Recognize and Support this Leadership," *One Earth* 4, no. 7 (2021): 907-909, https://www.sciencedirect.com/science/article/pii/S2590332221003572

36. Caroline Hickman et al., "Climate Anxiety in Children and Young People and Their Beliefs about Government Responses to Climate Change: A Global Survey," *The Lancet* 5, no. 12 (December 2021): e863-e873, doi:10.1016/S2542-5196(21)00278-3.

CHAPTER 12

Climate Change and the School Library

Ewa Dziedzic-Elliott

School libraries provide access to materials curated by certified and highly qualified school librarians for all students. In this chapter, I will share ideas and strategies regarding school librarians' ability to support students, educators, and school administrators in their effort to provide adequate education on climate change. This chapter can be used as an inspiration by school librarians or classroom teachers.

When searching for existing academic writings on school librarianship and climate change, I observed that most of the climate change and library articles focus on public librarianship, not school. I also found that many of the articles written about climate change and libraries are from other countries, especially Europe. My guess is that, in the United States, climate change is highly politicized and still considered to be controversial, therefore difficult to approach in some communities across the country. In New Jersey, climate change has been incorporated into all content areas in New Jersey Student Learning Standards (NJSLS),[1] which means that New Jersey educators are obligated to include climate change in their educational offerings.

Librarianship and Climate Change

Since 2016, the International Federation of Library Associations and Institutions (IFLA) has awarded prizes in two categories, Best Green Library and Best Green Library Project,[2] bringing international attention to the subjects of climate change, sustainability, ecology, and librarianship. In 2023, the winning library was Biblioteca EPM in Cucuta, Colombia, which placed sustainability at the heart of their strategies and implemented educational programs in sustainability.

The winning library program for 2023 was Climate Writer in Residence at the West Vancouver Memorial Library in British Columbia, Canada.[3]

The American Library Association (ALA) has guided librarians through the complex issues of climate change and sustainability and the role libraries can play in their local communities for decades. On January 10, 1990, ALA issued a resolution on the environment, urging librarians and library governing boards to collect and provide information on the condition of our earth and its air, ground, water, and living organisms.[4] In the spring of 2022, ALA released a document called *Sustainability in Libraries: A Call to Action*[5] laying out the basics of how library programs across the country can support efforts to create sustainable communities.

INFORMATION LITERACY AND CLIMATE CHANGE

Climate change can be a great way to incorporate Information Literacy (IL) into your practice. Information Literacy stands for understanding the need to seek information, locate the information resources, their evaluation, comprehension of what and why is being communicated and finally adding the component of appropriate information sharing. Too often we focus on the final outcome (writing a paper, preparing the presentation) and miss on opportunities to apply critical-thinking and problem-solving skills. Information Literacy forces us to ask the questions: Why and who is sharing this information? What do they accomplish by "selling" their perspective? Do I believe what I see just because it is published in print or online? In their research paper Boyer and Dziedzic-Elliott show that we often focus on mechanics of research, teach students how to access and navigate print and digital tools but don't challenge them to ask those crucial questions forcing them to critically evaluate the information in front of them.[6] Authors show that we get stuck focusing on information management level of research and are not being given a chance to explore critical thinking or metacognition of research.[7]

SCHOOL LIBRARIANS: A DIFFERENT TYPE OF EDUCATOR

School librarians are a unique type of educator, often holding multiple education certifications with past work experiences as teachers and librarians in other library settings, such as public or academic libraries. Unfortunately, not every school building has a school library or certified school librarian. According to a SLIDE study, over 20 percent of schools in the United States do not have school librarians.[8] New Jersey is following that unfortunate trend, as well and in some communities, public libraries are forced to support school-aged children in the absence of school library programs.

PROFESSIONAL TRAINING, LIBRARY PATRONS, SPACE, AND PROGRAMMING

School librarians are taught to work with state student learning standards, standards from the American Association of School Librarians (AASL), school districts' curricula, and local policies and regulations. They are also trained to collaborate with teachers in their building to support learning processes in the classroom. School librarians participate in professional development events that enhance their level of preparedness to create new learning environments, follow education trends, introduce new technology to their colleagues, provide various levels of tech support, and bring innovative teaching strategies to their buildings.

Climate Change and School Libraries

Stevenson, Nicholls, and Whitehouse call for climate change education to be extended outside of traditional structures and formal curriculum spaces to draw on new informal and hybrid (e.g., school/community) spaces and offer alternative possibilities for learning and action.[9] They confirm what school librarians strive to do by becoming an integral part of education outside the classroom.

At the core of any school library program stands a library collection made up of print and digital resources. Pötsönen et al. write that in Finland, libraries are considered environmentally friendly institutions due to their primary function of providing access to shared materials.[10] They look at the library's actions through the lens of passive and active tiers of library sustainability, such as being up to date with materials a library provides or involving the local community in their initiatives. The library also acts as a cultural hub with access to materials in other languages appropriate to local communities. Pötsönen et al. name four pillars of sustainability: ecological/environmental, social, cultural, and economic, all of which provide free equitably distributed access to professionally evaluated and curated information.[11] Finnish libraries are also known for access to a Library of Things (LoT), where patrons can borrow sports equipment, event tickets, devices, tools, and even bikes or moving carts. The authors point to support for education through library collection management, instruction on information literacy, and library events in collaboration with other organizations.

The LoT concept on its own is very climate-friendly: instead of purchasing and creating new things and wasting precious natural resources, we reuse the same ones. Of course, popular books in elementary schools will not survive as many circulation cycles as they might in middle or high school so they might need to be replaced more often.

SCHOOL LIBRARY CLIMATE CHANGE FRIENDLY COLLECTIONS

Building a collection that includes materials focused on climate change should be one of today's librarians' top priorities. Climate-change collections should include materials that meet the following recommendations:

1. Materials are up to date. Environmental science is changing and evolving, so provided resources must not be outdated.[12] The recommended age for most materials in fast-changing fields is about five years. In some cases, we could stretch this time frame to ten, but anything older needs to be closely evaluated for accuracy.
2. Make sure materials are in good condition. Students are more likely to check out books that are in good shape, don't have ripped out or stained pages, and smell good.[13]
3. Provide factual, unbiased information about the environment. Certified school librarians have tools that help them evaluate materials based on the quality of information.
4. Complement print materials with digital ones. These days many school libraries offer access to their collections on their library websites, including digital library catalogs, databases, and lists of properly evaluated and recommended internet resources. Databases are very expensive, so many state libraries provide free access to digital tools for state residents. In New Jersey, we have a tool called Jersey Clicks.[14] The current offer includes Primary Search for elementary school-aged students,[15] Middle Search Plus for grades 6–8,[16] a Green File database that specializes in environmental science for teens, Academic Search Premier with information in many disciplines appropriate for teens and young adults, and Points of View, a database that offers pro and con views of controversial issues.
5. Offer makerspaces. Proper makerspaces are tailored to the needs of the local community filling in the gaps in the absence of certain educational needs, for example, robotics, engineering, or computer programming classes. Those makerspaces are created to boost students' creativity and give them space to rest and reset. They might include board games, coloring tools, electronic or computer programming kits, and so much more. The most important part about creating these spaces is to make them local so they can truly serve students. Don't spend several thousands of dollars on a new 3D printer if there are already three in the building, but consider getting a box of wooden blocks

if students want to practice building environmentally friendly architectural designs.
6. Follow the Finnish LoT concept of including in collections nontraditional materials that serve local community, such as flower seeds, gardening tools, event tickets, sewing equipment, crocheting supplies, power tools, sports equipment, devices, and even bikes.[17] In economically challenged areas there might be a need to have education supplies available for our students, such as calculators, arts and crafts materials, headphones, digital devices, etc.
7. Environmentally friendly weeding. Weeding is the process of removing physical materials from the library that don't follow/support state standards or a district's curricula, contain factually incorrect or outdated information, have been damaged, the library has multiple copies with newer editions available, and so forth. It is recommended that libraries create policies that specify what can be done with weeded materials, for example removed books might be used for arts and crafts and upcycling projects. At one of my libraries, I partnered with an art teacher and used old books to build tables.
8. Extend your offerings outside the school. The library can serve as a cultural hub that offers community initiatives and materials in other languages appropriate for the local community.[18]
9. Brag about what is happening in the library. Sonkkanen, who writes about Finnish libraries and their state of sustainability, notes the need for appropriate dissemination of information about the work being done in the library.[19] She also points out the need to work with other organizations; collect feedback from library patrons; share best practices; measure and report results; cooperate with other libraries, library patrons, and organizations outside the library system; use the advantages of the library systems as a tool for communication; and add educational programs for children and youth.[20]
10. Create partnerships with other organizations outside of school libraries. Bringing engaging speakers, role models, or influencers can inspire students to follow their own pursuits. Authors who point out the need for such partnerships include Devine and Appleton,[21] Beutelspacher, and Meschede.[22]

School Librarians and Teachers Collaborations

School librarians are trained to find the best educational resources to support classroom teachers, students, and the whole school community. They know how to support students in each grade level with appropriate resources, traditional print and digital ones, and nontraditional. Think of the school library as the biggest classroom in the school available for schoolwide projects. The subject of climate change can be a great opportunity for a school librarian-teacher collaboration. Librarians can take on the role of research guides assisting students with finding credible information, while teachers can focus on writing style, composition of work, and assessing students' progress. Below, you will find a few examples of collaborative projects.

SCHOOLWIDE ACTIVITIES AND COMPETITIONS AND ACADEMIC RESEARCH

Competitions and contests are a great way to make research projects fun. They can be schoolwide or limited to a specific age group or grade level. It can become an annual tradition for the school to have an Earth Day door-decorating contest, the best poster of climate change or bulletin board. Every year can have a different theme, for example: climate change and our school, climate change and water, climate change and our town, etc. When working with teachers on research projects, we would often design the project together. The teacher would set their educational expectations and I would set up the guidelines for the research. Usually, the teacher would initiate the project in the classroom, and I would step in when they were ready to add their research component. It's important to consider the following questions when planning a project like this: What kind of resources would be most appropriate for student learning? Will we look at print or digital materials? How will we evaluate which websites provide real information on climate change? What are the red flags? What sources are most credible? Incorporating research allows students to apply their critical thinking skills and question the given information.

CLIMATE CHANGE COUNCIL

Create a Climate Change Council in the building and propose school/district policy changes. The program would have to start by evaluating the current environment through the lens of climate change and sustainability. We have state, national and international programs that offer assistance and partnerships providing tools, resources and guidance.[23] There is an opportunity for students to learn the school's operations, what and how gets approved, and by whom. School librarians can help teachers and students find the best resources to support climate change actions, including finding other school districts that share their goals. Librarians can assist in finding inspiring and replicable programs, even some from other countries.[24] Our students think outside the box and look at issues from their own perspectives, so their voices should be heard. The Council can start small, for example by challenging the plastic utensils and packaging in the school's cafeteria, creating pollinator gardens, and composting. All these ideas should have a research component. To make the right decisions or recommendations, students would have to research and learn what is the best replacement for plastic utensils in the cafeteria. What are the costs of the replacement? Are the materials manufactured in a sustainable fashion? Are they made locally? What are the native plants in our area? How do they need to be cared for? What kind of insects or animals will they attract? How to compost food waste? Where should it be placed in the school?

Recommended Websites and Internet Resources

Climate change is often targeted online by climate change deniers and conspiracy theorists. Looking for quality information on this subject might be very challenging. Librarians can co-teach classes that address the issue of climate change and the reliability of the resources, as well as provide tools that help students verify information. When I worked as a high-school librarian, I received a request from a teacher to create a list of reliable resources specific to our geographical area. We have a reservation of natural land in our state and the teacher wanted up-to-date information about it. We had some books, but we wanted to see what kind of credible sources were available online. After hours on the internet, I was able to put together a list of reliable resources that were used in the classroom. There are several websites that offer fact checking, such as PolitiFact, AllSides, Snopes, or FactCheck.[25]

New Publications

After teaching for a while, we all develop routines and strategies that work for us, but it is necessary to reevaluate them frequently. Even if you have taught a well-designed class on climate change for a long time, review the dates of the materials you are using. Are they older than five years? Maybe ten? If so, it's time to change them to something newer. Check in with your librarian and ask if any materials have been recently added to the school library collection. Have you learned about new materials you would like to see added? Make a suggestion. If there is no money in the school's budget for new resources, find out if there are local grant opportunities that can help. Grants can cover expansion of the physical collection or digital tools.

Makerspaces

Another interesting way for teachers to collaborate with librarians regarding climate change is by creating makerspaces. Makerspaces are programs that close the existing educational gaps in our classrooms. If there is no program that addresses robotics, arts and crafts, or engineering, libraries include those components in their offerings. When I became a high-school librarian, I wanted to add some of those ideas to my programs, but my students didn't want any high-tech tools. They said they needed to relax, take a break, and use their hands. They requested puzzles, coloring materials, Legos, and Keva planks.[26] Keva planks became very popular among both students and teachers. One thousand pieces of Keva planks became a tool to build various architectural designs and hold contests for who could build the highest tower. The planks also provided teachers a way to explain geometry or physics. Old-fashioned wooden planks became the center of the library. Again, if there is no budget for the purchase, write a grant with your librarian.

Climate Change, Upcycling, and Art

One of my favorite activities that can be done in collaboration between librarians and classroom teachers is upcycling random materials into artistic creations. Collect a variety of objects: old photo frames, buttons, string and twine, sand, rocks, shells, interesting jars or bottles. Add some paper and glue and let the children use their imagination. Use this as a teaching moment to show them we don't have to toss in the garbage things we no longer need. We can try to repurpose them, and that includes our clothes. There are a couple of different picture books where someone takes a coat and when it gets worn out, it turns it into a jacket, then a vest, then a handkerchief.[27] Those picture books can become the starting point of an intersecting climate change-oriented conversation: What should we do with unwanted clothes and household objects?

In the Absence of a School Library . . .

In the absence of a school library, partner with your local public library and/or academic libraries. Public libraries carry programs for children and young adults and often employ someone who fulfills the duties of a children or youth librarian. Although that person might not have excessive knowledge in state standards or your district's curriculum, they know a lot about literacy and have been trained to provide developmentally appropriate materials for school-age readers and learners. There might be one more advantage in working with a public library system: Their collection will have leisure-reading materials that the school library might not have the money for since school libraries mostly focus on academic support. In some school districts, partnerships between the school and public libraries are very close. They may hold joint events and initiatives, such as Pajama Storytime or an announcement of the summer reading theme. Public libraries provide access to art and entertainment programs, such as upcycling contests, movie screenings, workshops, and public readings. They also offer school visits, so you can invite your public librarians into your classroom and have them show off one of their programs. Younger students love a puppet show or arts and crafts based on a book.

Academic libraries are making the effort to provide much-needed support for school districts in many different forms, such as through school visits, or by creating lists of digital resources and offering training opportunities for school librarians and teachers. Some of these initiatives come from individual partnerships between school districts and higher education institutions while in others they have become systemic partnerships between academic and school library associations. Academic librarians specialize in various content areas and can become valuable members of the education community to provide professional development on recent climate change research, materials for K–12 schools, and digital tools.[28]

Conclusion

Whether struggling to find classroom materials to support your lesson plans or trying to create new and engaging ways to teach, please consider consulting your school librarian. If there is one in your building or district, maybe they can co-teach with you, or maybe they can provide you with a list of helpful professional reading materials or digital teaching tools. If the school library is not an option, reach out to the public library system. Is there a children's or youth librarian who can help you inspire your students to read? And there are always outreach or education librarians in academic institutions who are very knowledgeable and ready to provide training for you and your colleagues. Librarians can help.

Notes

1. "New Jersey Student Learning Standards," New Jersey Department of Education, January 31, 2024, https://www.nj.gov/education/standards/.
2. "IFLA Green Library Award," International Federation of Library Associations and Institutions, https://www.ifla.org/g/environment-sustainability-and-libraries/ifla-green-library-award/.
3. "IFLA Green Library Award."
4. American Library Association, *Resolution on the Environment* (Chicago: January 10, 1990), https://alair.ala.org/handle/11213/1574.

5. American Library Association, *Sustainability in Libraries: A Call to Action* (Chicago: April 2022), https://www.ala.org/aboutala/sites/ala.org.aboutala/files/content/SustainabilityInLibraries_Briefing_Final_April2022.pdf

6. Boyer, Brenda and Ewa Dziedzic-Elliott. "What I had, what I Needed: First-Year Students Reflect on how their High School Experience Prepared them for College Research." *The Journal of Academic Librarianship* 49, no. 4 (2023): 102742, https://www.sciencedirect.com/science/article/pii/S0099133323000812.

7. Ibid.

8. Keith Curry Lance and Debra E. Kachel, *Perspectives on School Librarian Employment in the United States, 2009-10 to 2018-19*. SLIDE: The School Librarian Investigation—Decline or Evolution? (Seattle & Washington, DC: Antioch University Seattle & Institute of Museum and Library Services, July 2021).

9. Robert B. Stevenson, Jennifer Nicholls, and Hilary Whitehouse, "What is Climate Change Education?" *Curriculum Perspectives* 37 no. 1 (April 2017): 67–71, doi:10.1007/s41297-017-0015-9.

10. Ulla Pötsönen, Leila Sonkkanen, and Harri Sahavirta, "Steppingstones to More Sustainable Public Libraries in Finland: From Individual Initiatives Toward National Guidelines and Standards," *International Journal of Librarianship* 5: no. 2 (2020), doi:10.23974/ijol.2020.vol5.2.179.

11. Pötsönen et al., "Steppingstones."

12. Ibid.

13. If the library collection is not maintained appropriately there might be a need to do a one-time evaluation of the collection. Removal (weeding) of damaged and outdated materials might be necessary.

14. Jersey Clicks, https://www.njstatelib.org/services_for_libraries/statewide_services/jerseyclicks/#1.

15. I received over 1,200 search results using the simple search term of "climate change."

16. I received over 6,800 results when I searched the term "climate change."

17. Pötsönen et al., "Steppingstones."

18. Ibid.

19. Leila Sonkkanen, "Sustainability Hides in Libraries: The State of Ecological Sustainability in Libraries," in *The Green Library – Die grüne Bibliothek* (Humboldt-Universität zu Berlin, Philosophische Fakultät I, Institut für Bibliotheks- und Informationswissenschaft, 2015), doi:10.1515/9783110309720.123.

20. Sonkkanen, "Sustainability Hides."

21. Jennie Devine and Leo Appleton, "Environmental Education in Public Libraries," *Library Management* 44: no. 1/2 (March 2023): 152-165, doi:10.1108/LM-10-2022-0091.

22. Lisa Beutelspacher and Christine Meschede, "Libraries as Promoters of Environmental Sustainability: Collections, Tools and Events," *IFLA Journal* 46: no. 4 (December 2020): 347-358, doi:10.1177/0340035220912513.

23. See: Sustainable Jersey Schools: https://www.sustainablejerseyschools.com/

24. See: Eco Schools Global: https://www.ecoschools.global/how-does-it-work

25. Politifact: https://www.politifact.com/; AllSides: https://www.allsides.com/unbiased-balanced-news; Snopes: https://www.snopes.com/; FactCheck: https://www.factcheck.org/.

26. https://www.kevaplanks.com/.

27. Jim Aylesworth, *My Grandfather's Coat* (Jefferson City, MO: Scholastic Press, 2014); Phoebe Gilman, *Something from Nothing* (Jefferson City, MO: Scholastic Press, 1998).

28. Gary Marks Jr., Grimes, N., & Lafazan, B. (2023). Academic and School Library Partnerships: An Organization-Led Collaboration. In *Cases on Establishing Effective Collaborations in Academic Libraries* (pp. 46-67). IGI Global.

CHAPTER 13

Sustainability and Climate Change for All

A SYSTEMATIC APPROACH TO IMPLEMENTING K–12 PROBLEM-BASED LEARNING AND DESIGN THINKING

Eddie Cohen and Brielle Kociolek

Climate change is one of the most pressing issues facing current and future generations. As global temperatures rise, sea levels increase, and extreme weather events become more frequent, finding solutions to mitigate and adapt to climate change has become critical. Schools play a vital role in educating and empowering students to understand and develop innovative solutions for climate change. This chapter provides guidance on leveraging two pedagogical approaches—problem-based learning (PBL) and design thinking—to engage K–12 students in tackling real-world climate-change challenges. This chapter offers resources and recommendations for educators to scaffold active investigation of climate issues in developmentally appropriate ways for elementary, middle, and high school students. With practice in PBL and design thinking, students can gain confidence in their ability to drive positive change.

Using Problem-Based Learning and Design Thinking to Solve Environmental Problems and Climate Change

PBL and design thinking are inductive teaching methodologies that empower students to take the lead in their learning journey. In 2020, the New Jersey Department of Education (NJDOE) adopted design thinking student learning standards for K–12 with the intention of preparing students to think critically and solve problems in an ever-changing technological society.[1] For the purposes of developing the PBL and design thinking climate change curriculum, the

Center for Mathematics, Science, and Computer Education (CMSCE) at Rutgers University used the following definitions:

- Problem based learning is a sub-branch of project-based learning, a student-centered approach in which the project is centered on a problem.[2]
- Design thinking represents a variety of concepts and skills that are relevant to any form of creative endeavor, especially in the context of real-world problem-solving and engineering design.[3,4]

These pedagogical choices are characterized by their focus on open-ended, real-world problems that serve as the core foundation for learning. Such problems are not only the catalyst for acquiring knowledge and skills, but they also provide relevant context. In these settings, teachers adopt the role of facilitators, guiding the learning process rather than directly imparting information. This shift allows students to engage in self-directed research and learning that are decentralized and collaborative.

A key aspect of these methods is that success is not judged solely on the solution found, but more importantly, by the depth and quality of critical thinking and inquiry students exhibit. They emphasize the importance of the learning process itself, encompassing analysis, synthesis, evaluation, and creativity. As such, the problems presented must be inherently multidisciplinary, encouraging students to draw connections across different subjects and see learning as an integrated experience rather than as isolated subject silos. For example, students must take a multi-modal approach to problem-solving when looking at air quality after wildfire smoke sweeps across numerous states or countries. Research becomes cross-curricular when they examine drier conditions related to climate change, forestry practices that vary across jurisdiction based on economic need and human interest, and patterns and trends across multiple datasets. Once causes are identified, students draw on their science and humanities backgrounds to suggest improvements in building codes that regulate housing in wildfire prone areas, a foray into psychology to improve human behavior, and even marketing that include everyone's favorite, Smokey the Bear that marches down NYC streets during Thanksgiving parades to remind us that "You can prevent forest fires!"

To drive this type of instruction, PBL projects must start with a problem that can be examined in multiple ways and that has various solutions not easily posed by students in a day or AI in seconds. Teachers continually need to identify meaningful real-world challenges without simple answers. In the PBL project "Human Impact and Understanding Sustainability: Energy" developed by Rutgers CMSCE, the problem posed to high-school students is, "Using the engineering design process you have been tasked by the NJ Department of Clean Energy to educate your community on how to help in transitioning to a more energy efficient community that can save your fellow citizens money and the environment." This problem allows students to research a multitude of energy solutions available in their specific areas and communities while considering cost/benefit ratios, including environmental, societal, and government (ESG) and equity. Solutions cannot be achieved by individuals on their own, because governmental regulations and shareholders demand financial success from the energy industry in a world where greenhouse gases cause climate change.

PBL and design thinking require collaboration in which students work together to identify what they already know, what they need to learn, and how to access the necessary information. Such collaboration supports them through the evolving process of problem-solving. Moreover, these methods encourage trial and error, viewing failures not as setbacks but as valuable opportunities for reflection and improvement. Hands-on experimentation becomes a key part of researching and testing potential solutions. Thus, these concepts provide students with the skills to develop and maintain a thirst for continuous learning, which is key to long-term success in post-graduation working environments.

Another crucial component of these teaching methods is documenting and communicating the learning journey. Encouraging students to articulate their processes and solutions can foster metacognitive awareness and help them engage in public reflection. Chapter 9 details key elements of using data visualization to enhance climate change education, which supports the documentation and communication students use throughout the PBL and design process. These elements enhance their understanding of all aspects of the investigated issues and help them develop important communication skills. We advocate for students to use engineering notebooks to document their design and engineering steps. Allowing students to maintain consistency across the various PBL projects they do across grade levels and subject areas promotes horizontal and vertical articulation. It also helps to facilitate collaborations across teacher teams in which students all use the same framework in PBL projects. When working with districts to support PBL implementation and adoption, it is essential to make knowledge-transfer explicit and easier for students by using the same framework! Overall, PBL and design thinking offer a dynamic and interactive approach to learning, placing students at the center of their educational experience.

Connecting Climate Change with PBL and Design Thinking

The increasing focus on climate change in formal education standards, such as in New Jersey's 2020 update to teaching and learning standards, which explicitly requires teaching climate change across seven K–12 subject areas, presents an ideal opportunity to integrate PBL and design thinking into the curriculum. These standards span core ideas in science, math, history, art, and engineering design. This breadth aligns perfectly with the multidisciplinary nature of PBL and design thinking, where students engage in learning that cuts across different subject areas.

Incorporating climate change into education can meet these new standards and leverage the critical thinking and civic engagement components at the heart of PBL and design thinking. New Jersey's Student Learning Standards were created to support climate change learning by requiring students to obtain, evaluate, and communicate information about climate impacts and mitigation strategies. For example, middle-school social studies New Jersey Student Learning Standard (NJSLS) 6.3.8CivicsPR.4 requires students to use climate data and evidence to communicate a proposal or defend a climate change-related policy.[5] This requirement aligns with the PBL and design thinking emphasis on self-directed inquiry and decentralized learning, where students actively seek out and analyze information.

Furthermore, the NJSLS supports climate change learning by encouraging students to construct explanations and design solutions supported by scientific evidence. This requirement dovetails with the PBL and design thinking focuses on students being in the driver's seat, formulating their own solutions to real-world problems. The learning process of analyzing, synthesizing, evaluating, and creating is a hallmark of these inductive teaching methods.

The standards also promote evidence-based debate around the advantages and disadvantages of various proposals to address climate challenges. Such an approach is intrinsic to PBL and design thinking, which encourage trial and error, hands-on experimentation, and collaborative problem-solving. This helps students understand the complexity of climate issues and fosters critical thinking and public reflection, essential components of civic engagement and scientific literacy. For instance, one possible solution for dealing with sea-level rise is increasing flood funds, and the community advocates through voting on those priorities. New Jersey sea-level rise is twice the US average and 50 percent higher than New York City's.[6] Should federal funding be allocated to states with higher sea-level rise then its peers? Once funds are allocated, how should they be used? Are sea walls the solution, or is a more natural rebuilding of marshland a better choice for protection? Towns along the coast now recycle Christmas trees by placing them into sand dunes along the shore, which creates faster sand deposits and strength as plants grow to stabilize new dunes. This solution was developed through trial-and-error during the 1990s and resulted in stronger dunes along Bradley Beach. These dunes prevented widespread damage after Hurricane Sandy. This solution was shared with others, and multiple states now use it to improve their own local dune ecosystems.[7]

In summary, the integration of climate change education with PBL and design thinking creates a dynamic and relevant learning environment. It adheres to the latest education standards, while equipping students with the skills and knowledge to tackle real-world challenges. This approach encourages active learning, fosters interdisciplinary connections, and develops critical thinking and problem-solving skills, preparing students to be informed and engaged citizens in addressing the pressing issue of climate change.

Helping K–12 students Embrace the Design Process to Develop Potential Solutions to Climate Issues

Supporting K–12 students in embracing the design process to develop potential solutions for climate issues requires a holistic approach that requires engaging schools, district leaders, administrators, supervisors, and teachers. This approach is crucial for effectively integrating PBL and design challenges into climate-change education.

Late school-year work at a suburban New Jersey Middle School provides an inspiring example of this approach. On the next to last day of school, one-hundred students gathered in small groups in their cafeteria for a design challenge to design wind-turbine blades (see figure 13.1).

This was a special activity for their transition day, where students in the lower school get together as a culmination of their sixth-grade experience and celebrate moving up to the upper school to begin as seventh graders in the fall. The activity offered a practical application of their learning and a fun and interactive challenge. Students measured the electricity generated by their wind turbines and then re-engineered their designs to create even more power. This iterative process sits at the heart of design thinking, encouraging experimentation and problem-solving that sparks new solutions to make improvements.

An important aspect of the NJ Middle School project was the carousel walk (see figure 13.2), where students examined their peers' designs and reflected on the various design choices.

This activity established individual small-group dynamics that contributed to broader group-think dynamics, fostering a sense of community and collaboration while allowing students to learn from each other to think more critically about their own designs. Teachers across multiple subject areas taught students aspects of the project that related to their subjects so when students came together in new groups, they could share their prior knowledge and collaborate to solve the problem at hand. For example, the math teacher asked students

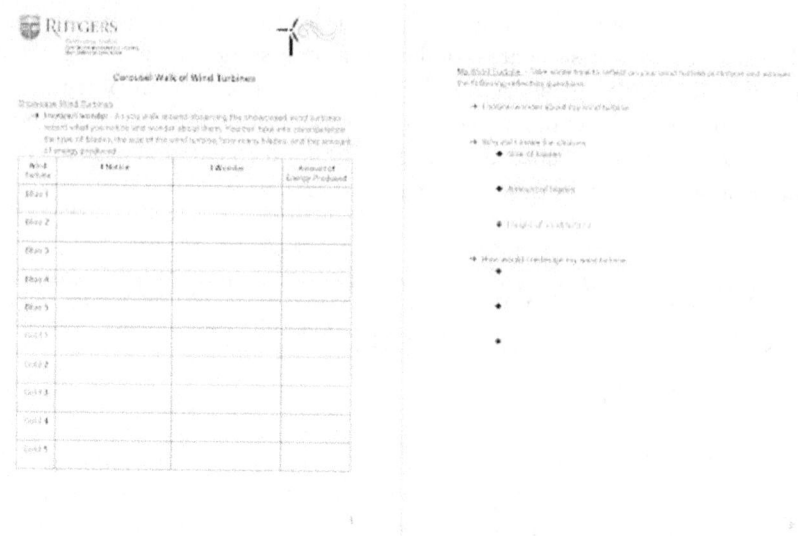

Figure 13.1 Design challenge guidelines for students.

Figure 13.2 Sample student wind turbine projects displayed during a carousel walk. E. Cohen

to examine fan blades and ratios of blades to power generation so they could decide whether to use two, three, four, or five blades. The art teacher worked with students to think about blade shapes. When these students came together to form strategically designed groups, they met new classmates in a meaningful collaborative way. This was a tenet of the school's goals for the transition day. The project culminated in a small competition, with prizes for the top three groups, adding an element of excitement and recognition to the learning experience.

The NJ Middle School example illustrates how educators can support students in developing metacognitive skills for self-assessment, progress-tracking, and ownership of learning through self-directed PBL. In this project, teachers played a crucial role in the setup process by working closely with students before they began designing their prototypes. Initially, teachers facilitated brainstorming sessions where students could share and discuss their ideas. This collaborative environment encouraged students to think creatively and consider various approaches to solving the problem at hand. Teachers guided these discussions, helping students refine their ideas and think critically about the feasibility and potential challenges of their designs.

Once the brainstorming phase was complete, teachers helped students plan their prototypes. They provided resources and demonstrated how to use simple tools and materials, such as small motors/generators, cardboard, glue, cork bottle tops, plastic spoons, and other lightweight items for building blades. Teachers ensured students understood the construction process and safety precautions when using tools like scissors and hand drills. Transforming the cafeteria into a design space before lunch showcased the project's practicality and ease of implementation. This temporary makerspace allowed students to work on their prototypes in a supportive, collaborative environment. The project was not only educationally valuable but also cost-effective, demonstrating that such initiatives can be easily replicated in other districts. The primary costs were minimal, involving inexpensive materials like those mentioned above. The low cost and ease of sourcing materials and tools make this project accessible for schools with limited budgets.

By providing resources, facilitating equitable learning environments, and guiding reflective meaning-making (helping students make sense of what they have learned by connecting new knowledge with their prior experiences with and beliefs about complex issues), educators play a vital role in shaping how students approach and solve real-world problems. Through such hands-on, engaging projects, students can learn about climate science and renewable energy and develop important skills in teamwork, critical thinking, and innovation. This approach helps students understand complex topics deeply and prepares them to tackle future challenges effectively.

Implementing PBL Across a School with Effective Horizontal and Vertical Articulation

Implementing PBL across a school with effective horizontal and vertical articulation is a multifaceted process that requires careful planning, execution, and reflection. The goal is to create a learning environment where students are actively engaged in real-world problems and developing critical thinking, collaboration, and problem-solving skills. The following primary stages should be examined when transitioning to a PBL pedagogical approach.

PLANNING STAGE

The planning stage is pivotal in laying the foundation for successful PBL implementation. It involves crucial steps like the following:

- Stakeholder meetings: Engage with supervisors, administrators, and teachers to discuss how PBL will be integrated into the curriculum.
- Determining scope: Identify relevant grade levels and subject areas for PBL application.
- Setting timelines: Establish a feasible period for implementation.
- Professional development: Conduct sessions to orient teachers to PBL methodologies and address potential concerns.

IMPLEMENTATION STAGE

During the implementation phase, the following strategic actions are taken:

- Adopting a lesson study approach: Teachers collaboratively plan and observe research lessons, refining their approach based on feedback.
- Identifying opportunities: Utilize unique scheduling opportunities for PBL activities, such as the week before summer break or in place of previous team-building experiences that are already acceptable experiences for students.
- Resource organization: Gather necessary materials, including those in a makerspace, to support PBL activities.
- Observations and feedback: Facilitate rotating teacher observations to provide diverse insights and feedback for continuous improvement.

DEBRIEFING STAGE

The debriefing stage is crucial for reflection and future planning:

- Pre- and post-implementation meetings: Engage in discussions with administrators, supervisors, and staff to analyze what worked well and areas for improvement.
- Future planning: Create PBL/design units for each grade level and plan through professional learning communities. This can be done during traditional curriculum writing time periods.

ENHANCING THE PBL ENVIRONMENT

Several factors contribute to creating a supportive environment for PBL:

- Cultural shift: As Edutopia points out,[2,9] it's vital to create a culture that values the expertise of all teachers and integrates PBL with existing teaching practices.
- Content selection: Choose authentic, real-world connected projects that align with the curriculum.
- Student mindset: Gradually introduce students to PBL and increase task complexity over time.
- Inquiry-based learning: Scaffold the process of research and inquiry and encourage interdisciplinary collaboration.
- Classroom organization: Redesign spaces to facilitate collaboration and make materials accessible.

IMPROVING OUTCOMES AND ENGAGEMENT

PBL has shown considerable benefits in improving student outcomes and engagement:

- Enhanced learning: A randomized study found that students in a project-based version of Advanced Placement (AP) courses that used the Knowledge in Action program achieved a higher number of credit-qualifying scores on the AP tests than students who experienced traditional instruction—eight percentage points higher after one year and ten percentage points higher after two years.[8]

- Equity in education: PBL provides opportunities for all students to engage in culturally relevant community-focused projects, making education more equitable and inclusive.[9]

KEY CONSIDERATIONS

Several key considerations must be kept in mind when implementing PBL:

- Integration with existing practices: Rather than replacing existing methods, PBL should be integrated to complement and enhance the current curriculum.
- Teacher preparedness: Continuous professional development and support for teachers are crucial for effective PBL implementation.[10]
- Student readiness: Prepare students for a shift from traditional learning methods to a more inquiry-based, autonomous learning approach.
- Resource allocation: Ensure the necessary resources, both physical and digital, are available and accessible for students and teachers.

CONTINUOUS IMPROVEMENT

The process of implementing PBL is ongoing and requires continuous reflection and improvement:

- Feedback mechanisms: Regular feedback from students, teachers, and administrators is essential to refine the PBL approach.
- Adaptability: Be open to adapting PBL strategies based on feedback and changing educational needs.
- Monitoring progress: Regularly assess the impact of PBL on student learning outcomes and adjust as needed.

In conclusion, systematically implementing PBL in schools is a dynamic and comprehensive process that involves careful planning, execution, and ongoing refinement. It requires the collective effort of all stakeholders to create an engaging, effective, and equitable learning environment. By focusing on these key stages and considerations, schools can successfully and systematically integrate PBL into their educational framework, benefiting both students and educators.

Supporting and Training Teachers and Informal Educators to Create and Implement Climate Change PBLs

School and district leaders are pivotal in enhancing professional learning focused on climate education and best pedagogical practices. Organizations like CMSCE play a vital role in this endeavor by providing interactive workshops that model frameworks such as PBL. These workshops are designed to equip educators with the necessary skills and knowledge for designing and facilitating engaging, climate-focused learning experiences.

These workshops offer a range of flexible lessons and tools, each tailored to different educational levels from elementary school to high school. They address developmentally appropriate climate themes, ensuring the content is both relevant and accessible to students of varying ages. Educators are guided through a variety of activities in these workshops, including the following:

1. Conducting close readings of climate standards across grades and disciplines to recognize alignment with PBL goals.
2. Using school sustainability audits to ground abstract climate topics locally for place- based education.
3. Evaluating quality informational texts, simulations, and case studies spanning climate subject matter.
4. Curating grade-level playlists of online climate change modules for just-in-time student inquiry.
5. Developing student assessment rubrics tied to climate competencies like systems thinking, analyzing tradeoffs, interpreting models, and evidence-based policymaking.
6. Designing project arc outlines and facilitator guides to support student pace-setting and independence.

These resources and models provided by CMSCE workshops significantly lower the barriers for educators, boosting their confidence in integrating climate education across various subjects and setting their students up for success.

An exemplary instance of this approach is the Rutgers CMSCE Design Challenge/PBL on carbon sequestration. This project offers a practical application of these methodologies, demonstrating how they can be effectively utilized to engage students in critical thinking and problem-solving around key issues in climate change and sustainability. Through such initiatives, educators are not just imparting knowledge but are also empowering students to actively engage with and contribute to solutions for real-world environmental challenges.

Conclusion

Anchored in real-world contexts, climate change serves as a timely and profoundly important topic for cultivating systems thinking and civic agency through PBL and design thinking. Students investigate multidimensional causes and cascading consequences of climate disruptions across the atmosphere, biosphere, hydrosphere, and anthroposphere. In seeking potential interventions, they weigh tradeoffs around economics, ethics, technology, politics, and social justice. Throughout this integrated learning journey, students practice quantitative and qualitative reasoning, critical analysis, creative ideation, and communicating compelling evidence-based narratives.

With appropriate scaffolding around climate literacy, metacognitive skills, and collaborative inquiry, PBLs enable students to achieve academic standards while advancing their intellectual and empathetic maturity. Design thinking equips students to move from abstraction to invention around concrete climate problems impacting their communities. Through successive experimentation and debate, youth recognize their growing ability to drive change amid immense challenges. Such active, empowering education fuels students' resolve for lifelong climate stewardship and earnest civic participation. This better world awaits their ideas, passions, and persistence.

Notes

1. New Jersey Department of Education, "Computer Science & Design Thinking," https://www.nj.gov/education/standards/compsci/Index.shtml

2. John Larmer, "Project-Based Learning vs. Problem-Based Learning vs. X-BL," *Edutopia*, July 13, 2015, https://www.edutopia.org/blog/pbl-vs-pbl-vs-xbl-john-larmer.

3. The Interaction Design Foundation, "What Is Design Thinking?," February 6, 2024, https://www.interaction-design.org/literature/topics/design-thinking#:~:text=Design%20thinking%20is%20a%20non,%2C%20Ideate%2C%20Prototype%20and%20Test

4. Moinul Alam, "What is the Design Thinking? Definition, Importance, Examples, and Process," IdeaScale, April 18, 2024, https://ideascale.com/blog/what-is-the-design-thinking/

5. New Jersey Department of Education, "Social Studies," https://www.nj.gov/education/standards/socst/

6. James Shope et al., *State of the Climate: New Jersey 2022* (New Brunswick, NJ: Rutgers New Jersey Climate Change Resource Center, 2022), https://njclimateresourcecenter.rutgers.edu/wp-content/uploads/2023/04/State-of-the-Climate-2022-042423.pdf

7. Mireya Navarro and Rachel Nuwer, "Resisted for Blocking the View, Dunes Prove They Blunt Storms," *The New York Times*, December 4, 2012, https://www.nytimes.com/2012/12/04/science/earth/after-hurricane-sandy-dunes-prove-they-blunt-storms.html

8. Lucas Education Research, "Project-Based Learning Boosts Student Achievement in AP Courses," 2021.

9. Youki Terada, "New Research Makes a Powerful Case for PBL," *Edutopia*, February 21, 2021, https://www.edutopia.org/article/new-research-makes-powerful-case-pbl

10. Ibid.

CHAPTER 14

Incorporating Climate Emotions into the Classroom
STRATEGIES TO SUPPORT STUDENT EMOTIONAL GROWTH AND THRIVING

Kathleen L. Grant and Sarah Springer

Comprehensive climate-change education is critical for educating young citizens and future leaders about the complex scientific, social, political, and economic impacts of our changing planet. Educators must also consider the range of emotional responses children and adolescents may experience as they are exposed to new ideas and have space to process current thoughts and feelings around climate change. As progressive education acknowledges, teaching and learning are most effective when educators invite the whole child into the classroom: cognitively, physically, and emotionally.

Not only can teachers be more effective when learning is connected to feelings and emotions, but an increasing number of learners are also impacted by the climate crisis and bring their climate-related experiences, fears, traumas, anxieties, losses, and hopes into the classroom. More and more students will be exposed to extreme weather-related events associated with climate change, such as wildfires, floods, and extreme heat. Some students will have traumatic experiences related to these weather events and may exhibit trauma responses in the classroom, potentially more so when triggers are present. Additionally, a growing number of youth and adults are dealing with climate anxiety. Educators must be prepared for the range of emotional reactions students can exhibit when learning about the climate crisis and equipped with the skills to help students move through these feelings and reactions in an adaptive way.

This chapter aims to provide an overview of youth mental health in relation to the climate crisis, focusing on how climate emotions can be seen and addressed in the classroom. It will provide educators strategies to incorporate a deeper understanding of climate emotions into their professional practice, starting with their own emotional wellness and then providing strategies to work with young people and classroom communities. Finally, a vignette will illuminate how basic techniques can be incorporated into classroom discussions.

Overview of the Climate Crisis's Impact on Youth Mental Health

The American Psychological Association defines climate anxiety as a "chronic fear of environmental doom."[1] The associated feelings can include grief, fear, anger, worry, guilt, shame, and despair. Large-scale ($n=10,000$) studies have documented that most surveyed youth (seventy percent) believe their future is frightening, and almost half indicate that feelings about climate change impact their daily lives.[2] Youth are particularly vulnerable to climate anxiety for several reasons, particularly due to a lack of developed coping mechanisms to understand and process such fears and a real or perceived powerlessness to effect change.

While severe anxiety can negatively impact youth wellness and functioning, including the ability to learn and develop agency, climate anxiety is typically conceptualized as a legitimate response to a realistic threat. Anxiety can serve helpful purposes, especially when it motivates us to make changes and act. Climate anxiety may manifest in young people experiencing an existential fear for the future, wondering what the world will be like when they are adults, fearing there won't be safe places to live, and questioning whether to have children. They may also experience anger toward adults, feeling they have failed them and have not protected them from complex threats to a safe future.

While many youths experience fear for their future, some are also negatively impacted by the current effects of our changing climate. Youth are exposed to intense events such as wildfires, flooding, and extreme heat, which can lead to short- or long-term conditions such as trauma, anxiety, post-traumatic stress disorder (PTSD), depression, hopelessness, and aggression. For example, in a community that experienced catastrophic flooding, resulting in families and children being airlifted to safety via helicopter, some elementary school students experienced symptoms of PTSD (flashbacks, panic, crying) in their classrooms whenever it rained or when they heard emergency sirens. Research has demonstrated that children exposed to wildfires have increased rates of anxiety disorders, panic attacks, flashbacks, acute stress disorder, and psychotic disorders (citation).[3]

Eco-grief is the feeling of loss people experience by witnessing ecological changes, such as the loss of ecosystems, species, and personally important places and ways of life.[4] Youth may no longer be able to swim in a beloved river because it is too polluted or play outside because it is too hot, or the air quality is too poor. Eco-grief refers to strong feelings about our environment's current or future losses. Eco-grief is seen as both a risk factor for negative mental health outcomes and a potential source to motivate positive environmental behavior.

Youth with marginalized identities experience a disproportionate impact of the climate crisis. Low-income, Black, and Indigenous communities are often relegated to the least desirable locations, such as in flood plains; urban areas with a lack of essential tree canopies, drastically increasing rates of extreme heat; and downwind of fossil fuel-burning plants, contributing to increased rates of childhood asthma in Black youth over White youth. Black youth are eight times more likely to die of childhood asthma than White youth.[5] Young people in marginalized communities may be acutely aware of the impact environmental racism has on the health and wellness of their community members.

Research suggests that when caregivers are aware of children's emotional responses to eco-anxiety, children are more likely to build adaptive coping strategies.[6] It is likewise important to note that the impact of climate change on children's mental health can be felt through both direct exposure and vicarious experiences. This suggests that the media's representation of climate concerns and adults' responses to these impacts profoundly affect how children deal with their own feelings associated with climate change. Therefore, adults must recognize their emotional reactions to climate change to support the children and adolescents around them most effectively.

Learning about the climate crisis can trigger anxiety, fear, and hopelessness due to both the severity and uncertainty of climate change.[7] Climate change threatens humanity's survival and raises serious moral, existential, and spiritual questions (e.g., "Is our western way of life moral given its impact on the planet?") teachers and learners must grapple with as they engage with climate change education. These difficult questions can lead to feelings of fear and hopelessness, along with coping strategies such as avoidance and numbing to ease the pain and discomfort associated with challenging feelings. A critical emotional awareness is necessary to engage with the complexity of the climate crisis.

But I'm Not a Counselor!

Teachers are tasked with increasing responsibilities in schools. Attending to the climate-related emotional needs of students in their classrooms may seem like yet another task on the insurmountable to-do list. Additionally, teachers may feel they lack the training and expertise to address issues related to mental health in their classrooms competently. Some educators may believe their role is to teach "just the facts," and they are less comfortable with engaging the emotional factors that undergird certain types of learning.

It is critical to consider both points and recognize the barriers teachers may feel as they engage with climate-related emotions. This chapter does not suggest that teachers diagnose or treat any mental health conditions; rather, they can work to create space for students to bring climate-related emotions into the classroom. Connecting feelings to content can be beneficial when engaging with climate issues for several reasons, including the following:

- *Learning is more effective when connected with feelings and lived experiences.*
    Emotions can either enhance or interfere with learning. Emotions influence various aspects of cognition, such as critical thinking, memory, and attention.[8] Positive emotions also promote motivation and support the development of mastery-based goals.
- *Difficult feelings are associated with the climate crisis.* To be engaged in long-term action, youth need to learn how to effectively feel and manage their climate-related emotions, not suppress or deal with them maladaptively. The classroom can provide a space for students to learn how to acknowledge and use climate-related emotions as a source of knowing and even a method to drive action.
- *The climate crisis requires new ways of being.* Climate crisis education requires new ways of being with students and modeling the skills necessary to thrive in the future.
    Part of the solution to the climate crisis is acknowledging the ineffectiveness and harm rooted in current societal practices. The classroom can be a space to move toward new ways of being that center the whole child (their cognitive, emotional, physical, and spiritual selves), community, interconnectedness, and care.

Being an emotionally aware climate educator does not require special skills or training. It requires us to meet students where they are developmentally and holistically, recognizing that feelings are powerful sources of knowing in the classroom and can be welcomed as part of the whole child. When emotions or mental health issues overwhelm youth or get in the way of learning, teachers should consult with a school counselor for additional support.

To make space to address emotions in the classroom, teachers may need to assess which of their responsibilities must be deprioritized to create the time for this work. Educators should not continually be asked to take on more, sometimes at the expense of their own wellness, to endeavor to meet students' needs. Teachers must also center their wellness in this work. Educators can only meet students' emotional needs if their emotional needs are met. Additionally, teachers are important role models, as students watch them to emulate how to be in the world. Teachers have a powerful role to play in modeling ways of being, ways that give them the resources they need to address the massive challenges ahead.

Strategies for the Person-As-Educator

Emotionally aware climate educators provide youth with time and space to feel various feelings associated with the climate crisis and fears about the present and future. To create such spaces, teachers need resources. Most importantly, teachers need the emotional bandwidth to meet their students with difficult emotions like fear, grief, and sadness. While there is no one way to cultivate emotional capacity, some suggestions are provided below.

Journaling

Journaling is an effective way to increase awareness of the challenges teachers face. It can help the writer explore challenges, find insights, and process stressful situations. Journaling can help teachers recognize their own inclinations to lean into or move away from the "hard to feel" feelings that may come up in climate discussions, such as fear, worry, sadness, and shame. The more individuals can "make friends" with difficult feelings, the more able they will be to sit with the range of human emotions. One prompt is provided below:

- When you hear about wildfires in California, a drought in Colorado, a heatwave in Vermont, a flood in Texas, or a hurricane in New Orleans, what emotions come up for you? Which of the three coping strategies above are you most likely to employ? Is there a change in your emotional response when you are geographically closer to a climate concern? At what times in your life do you employ psychological distance from climate-related disasters?

COMMUNITY GROUPS/CIRCLES

Teaching about climate change can be emotionally challenging work. Community groups of educators can provide space for teachers to discuss the range of feelings in this work, as well as the joys and challenges that occur as they engage students in this material. By creating space to address their needs and support their colleagues, teachers will be more able to create spaces that hold their students' complex reactions. Climate Awakening provides an open, supportive, and confidential online space to process emotional responses to the climate crisis.[9] It can be a useful space for teachers to discuss their emotions related to teaching about climate change.

Additionally, connection with community is a practice that promotes positive mental health and is an essential strategy to bring about change when addressing the climate crisis. Complex problems can only be solved by individuals working in collaboration. Also, people engaged in action within a community are more likely to maintain their efforts over time, leading to more effective engagement. Schools are often rooted in individualism; therefore, educators can benefit from spaces where they can support each other in the important work of teaching climate content and gain essential support and care needed to promote healthy functioning and ongoing effective engagement in the classroom.

NATURE

Being out in nature is an effective way to deal with stress and climate anxiety. As little as ten to twenty minutes a day outside in a natural space like a forest can significantly reduce stress, fatigue, anxiety, and depression in both adults and youth.[10] Time in nature also promotes feelings of calm and positive affect. Engagement in nature can provide mental health benefits while cultivating a deep love for the natural world, which is also linked to a greater commitment to climate action. Educators can consider how they can incorporate natural experiences into their daily or weekly routines, such as walking in green spaces, watching the stars, tending to a garden, or listening to waves at the seashore.

REST

Educators, and all those who deeply care about others, cannot engage in this work when we do not take care of ourselves. However, the dominant American culture, informed by capitalist values, often prides itself on people pushing themselves past their limits. It promotes the false belief that one's worth as a human is linked to what one can produce. Many educators may feel trapped in this cycle, exhausted, and burnt out.

Climate-change education is predicated on the fact that unsustainable human norms and habits must be transformed to prevent the direst possible outcomes for life on earth. As we dream and discover new (or forgotten) ways of being in the world, rest can be a central and radical practice that helps us align ourselves with the changes that must occur and provides space for essential self-care. Slowing down and resting allows us to regain connection with ourselves in a system based on losing connection with ourselves, each other, and the natural world. Healing, change, and transformation can come from deep listening to our bodies and earth. Trisha Hersey's book *Rest is Resistance* is a field guide to a deeper practice in rest.

Youth Climate Emotions in the Classroom

Climate-aware educators acknowledge that their students may be experiencing a range of mental health issues and emotions in relation to the climate crisis. As educators increasingly address climate change in the classroom, they must consider how the content impacts students emotionally and how to help students learn, experience, and express their emotions effectively. Educators must keep this in mind when they are designing and implementing lessons and cultivating classroom communities.

Research suggests that teachers, consciously or not, create the emotional norms and rules within a classroom.[11] These rules dictate what is safe to share within the learning environment and what emotions are taken seriously. In one study, high school students who thought their teacher would not take their negative feelings about climate change seriously or would even make fun of their emotional expressions were more likely to minimize the seriousness of climate change.[12] Children and adolescents need a safe place to process their emotions and express their vulnerability. Youth are more likely to share within the context of trusting relationships, both with peers and teachers. Educators must acknowledge the seriousness of climate change and allow students time and space to articulate their worries.[13]

The most common emotional response to learning about climate change is worry, often mixed with feelings of guilt and/or hopelessness.[14] Worry can be a negative force, especially when no strategies exist to deal with the root cause of the worry. Youth may withdraw from cognitive and emotional engagement as a defense mechanism or more maladaptive forms of coping. However, worry can also be a positive force and spur people to action. Discomfort is inevitable when learning about climate change. While worry, anxiety, and dissonance are hard to bear, young citizens must develop the capacity to tolerate difficult feelings to stay engaged in learning and the possible individual and collective steps that can support change. Young people may also feel hopeless and frustrated, especially when realizing actions they are in control of as individuals are not enough to make the necessary changes to avoid the direst consequences of the climate crisis.

Typically, anxiety, worry, frustration, and hopelessness are considered "negative" emotions. However, viewing emotions as neither "good" nor "bad" but simply as normal responses is important. It may be helpful to reframe the terms "bad" feelings with "hard to feel" and "good" with "easy to feel," which may allow young people to accept all feelings without judgment. Each student may fall into particular emotional patterns when confronting difficult content or triggers, which teachers can help students increase their awareness of and examine how students cope with "hard to feel" feelings.

Youth have common coping responses to eco-anxiety. Students may engage in problem-focused coping, which typically reflects an individual's action-oriented response to an issue; emotion-focused coping, which is described as the awareness of negative emotions and the subsequent active steps to avoid, de-emphasize, or deny them; and meaning-focused coping, which similarly acknowledges the negative emotions but also employs strategies to reframe these emotions in a way that promotes action. Meaning-focused coping appears motivational and connected to hope and pro-environmental action.[15] Educators who can identify a variety of emotional reactions to climate change and provide spaces to help leverage this energy into an adaptive collective response may provide meaningful opportunities for youth activism.

RESILIENCE

Resilience is an individual's capacity to survive and thrive in the face of adversity. Resilient people are more able to deal with life's stressors effectively. As the challenges associated with climate change will increase significantly in the coming years, resilient youth will be more able to manage the stressors inherent in the complex, volatile, and unpredictable future. However, resilience is a social and emotional skill that can be modeled to youth and strengthened over time. Emotional awareness of self and others, connectedness, self-efficacy, and hope are all key factors contributing to the development of resilience. Incorporating resilience-building strategies in the classroom, such as setting brave goals, taking risks, labeling emotions, and recognizing that challenges are critical to success, can effectively support students' current and future emotional functioning.

However, the goal is not simply to increase young people's ability to endure hardship and suffering. Resilience used to deflect attention from oppressive and harmful systems is problematic. Educators must recognize the importance of working to prevent hardship from occurring. Resilience building can be used to address the unavoidable suffering that will occur because of climate change.

Dealing With Strong Student Emotions in the Classroom

Climate-aware educators recognize and welcome the wide range of emotional responses young people will have when learning about the climate crisis. They recognize feelings are natural for everyone and give us important insights into students' inner worlds. Feelings can also give us different ways of understanding unjust or difficult situations.[16] The following example is a reaction an eleventh-grade student, Maria, had to learning content that did not seem meaningful to her:

> What is the point of learning this? There is going to be no place for us to live. The west will be on fire, the southwest won't have water, and the coastlines will be flooded. What is the point of learning this meaningless math when my future will be disastrous?

Maria's strong emotions provide valuable insight into her experience and may suggest that she needs additional care and support. The following steps can be useful when working with strong student emotions in the classroom.

Step 1: Check in with yourself.
What feelings come up for you as you listen to Maria? Perhaps grief, anger, annoyance, or sadness? Allow yourself to feel and name the emotions that arise. What thoughts come to mind? You may find yourself thinking about young people in your life and what their future may hold. Focus on your breath, allowing it to ground you.

Step 2: Extend empathy.
Put yourself in Maria's shoes. What must it feel like to have such an uncertain future? What else might she be feeling? Try to find the nuance and complexity of Maria's experience and pain.

Step 3: Resist the urge to fix.
It is natural for many adults to wish to solve young people's problems, especially children's. Adults may desire to "make it better" for youth to protect and care for them. However, sometimes adults seek to "fix" as a strategy to avoid seeing the extent of young people's pain. The adult may not be able to tolerate the youth's big feelings. Also, jumping in to "fix" the problem undermines the child's ability to create an authentic solution for themselves. Finally, there are no easy solutions as we tackle the climate crisis; young people need to learn to tolerate the difficult feelings as they arise.

Step 4: Validate feelings and invite the youth to share more.
By validating the young person's emotions, educators signal that the emotions are real and important, which builds emotional awareness and regulation. Additionally, if educators can tolerate big emotions, students learn to feel a range of emotions and maintain their belonging in the classroom community. A simple refrain, "That makes sense, tell me more," will validate Maria's statement and invite her to keep exploring her experience and feelings. Students can also feel alone in their experiences, but the fears around climate change impact many of us. If this is the case for the teacher, she could respond, "The future you describe sounds really scary. I, too, can be really afraid for the future, for both me and my children."

Step 5: Embrace hope.
While teachers can embrace a range of "hard to feel" feelings in students, they do not want to end there, which can lead to feelings of hopelessness or despair. After educators validate students' feelings, they can comment on humans' profound

ability to create a new future that is healthy, safe, and more equitable. In one study, young people who received more hopeful messages about the climate crisis used more meaning-focused (reframing stress; finding hope and purpose) and problem-focused (addressing the source of stress) coping strategies rather than emotion-focused ones (avoiding or suppressing feelings).[17] The teacher could say, "Many parts of our future look bleak; I agree with you. But many good people are working hard to create solutions that will lead to a safe and healthy future. It is an exciting time to be alive, and we can all be a part of the work of transforming our world."

Classroom-Based Activities

Climate-aware educators can incorporate activities into the classroom that support students' understanding of their emotions related to climate change. By allowing climate-related emotions in the classroom, students may be better equipped to engage with climate content in the classroom and action in the broader world. Several strategies are discussed below.

CLIMATE FEELINGS CHECK-IN

Students can benefit from time to discuss the range of emotions they experience related to climate change. Educators can provide students with a nonjudgmental space to reflect on their feelings and how they may change over time. Teachers can encourage students to share a range of feelings, the "easy to feel" and "hard to feel" ones, and look for signs of hope. It is especially important for young people to be able to put their worries into words, gain control of the worry, and take action to deal with it.[18] The *Climate Doom to Messy Hope Handbook*[19] has specific suggestions and prompts for classroom discussions.

READINGS/GUEST LECTURES

Students can benefit from examples of people who channel their climate emotions into positive action. Educators can provide readings that highlight these individuals and consider inviting community members to class to discuss their journeys in environmental action and corresponding emotions.[20] Providing students with examples of people engaged in climate action, such as scientists, politicians, and community activists, can help them develop hope that people in their communities are working to bring about change and provide possible pathways for their current and future engagement.

COLLECTIVE ACTION

Action is a powerful tool to deal with "hard to feel" climate emotions, such as climate anxiety, worry, and hopelessness. Engagement to bring about change can also contribute to positive mental health outcomes, such as feelings of agency, self-efficacy, purpose, and connection. Students can work together on classroom projects to bring about collective change or engage with community groups doing such work.

MINDFULNESS

Mindfulness is increasingly used in mental health and education to support healthy youth functioning. Youth can benefit from expanding their ability to tolerate a wide range of emotions without becoming dysregulated or utilizing maladaptive coping mechanisms. There are many mindfulness activities, including breathing exercises, meditation, and yoga. However, any activity where students are fully present and actively aware of their thoughts and feelings can be a mindfulness practice. Teachers can incorporate practices that work best for their unique styles and their students' developmental stage. Possibilities include spending time in nature while engaging all our senses, taking regular breaks to check in with our breath, and working with students to develop practices that bring peace and calm (coloring, listening to soothing music, etc.). Educators can create mindful spaces that encourage presence, deep listening, compassion, and connection.

Vignette

Ms. Lopez is a fourth-grade teacher addressing New Jersey Student Learning Standards 6.1.5.GeoHE.3 (analyze the effects of catastrophic environmental and technological events on human settlements and migration). During a large-group instruction session, a student, John, jokingly grabs his friend and yells, "We are all going to die!" Upon seeing this, one of the girls, Lanie, starts hysterically crying. Ms. Lopez's initial reaction is anger toward John because she sees his reaction as inappropriate. He has disrupted her lesson and has activated his classmates. Ms. Lopez notices her feelings nonjudgmentally, taking a moment to breathe and ground herself. She remembers that children have a range of emotions when learning about potentially challenging content, and all emotions are okay and important, even if the behavior is not. She reconceptualizes both John's and Lanie's reactions as developmentally appropriate responses and recognizes they both may benefit from additional tools to verbalize their complex thoughts and feelings. She sees her role in this moment to help her students identify their feelings, verbalize them, and provide validation.

After grounding herself, Ms. Lopez says, "I'd like to stop right here and take a moment to acknowledge that there are a lot of feelings expressed in our community right now. Does anyone else notice this? Let's take a moment and notice how we are feeling right now." She then says, "I would like to invite any student experiencing a strong feeling to express it with our group, if they are comfortable." Lanie shares, "I am feeling really scared right now after John said we are all going to die!" Ms. Lopez thanks Lanie for her courage in sharing with the group and validates that it can be terrifying to consider we will die. She does not try to change Lanie's perspective or point her toward a more optimistic view of the future; she just allows Lanie to feel exactly what she is feeling now.

Ms. Lopez invites John to share his feelings. He states, "I am also really scared when we learn about the wildfires. My friend used to live in California and said there were days when the whole sky looked like it was on fire, and he had to stay inside. I didn't mean to upset Lanie; I am just really scared, too." Ms. Lopez expresses appreciation for John's sharing, and she validates how scary it must be to learn about his friend's experience and that she, too, has been frightened when seeing images of wildfires. She also notes that laughter, John's reaction, can be a common reaction to anxiety but could be interpreted by others as insensitivity. Ms. Lopez reflects on how she, John, and Lanie reacted differently to the presented content but had similar underlying feelings. She asks John and Lanie how they felt after they shared. Finally, Ms. Lopez states, "Climate change is creating frightening conditions worldwide, and it is important to recognize that and feel whatever feelings emerge. We must discuss our feelings with friends, teachers, and family. We may encounter hard challenges associated with climate change, but nothing is too hard for us to get through together. There are also people everywhere working to create change, and many exciting changes are happening every day. We can be part of the change, too."

Conclusion

Climate change education requires both teachers and students to consider emotions as part of the learning process. Increased emotional awareness and regulation are necessary when engaging with educational content regarding the complex and urgent challenges of the climate crisis and are essential for effective and sustained climate action. The classroom can be a critical location in which to support teachers and students in developing strategies to feel and articulate a wide range of emotions and have their reactions seen and validated by peers and teachers. Educators can continue to consider how to model classroom practices necessary for citizens to thrive and create change in an increasingly complex world.

Notes

1. Susan Clayton et al., "Mental Health and Our Changing Climate: Impacts, Implications, and Guidance," American Psychological Association. https://www.apa.org/news/press/releases/2017/03/mental-health-climate.pdf.

2. Caroline Hickman et al., "Climate Anxiety in Children and Young People and Their Beliefs about Government Responses to Climate Change: A Global Survey," *The Lancet* 5, no. 12 (2021): 863–73, doi:10.1016/S2542-5196(21)00278-3.

3. Medard Adu et al., "Children's Psychological Reactions to Wildfires: A Review. Current Psychiatry Reports 25, no. 11 (2023): 603–11, doi:10.1007/s11920-023-01451-7.

4. Hannah Comtesse et al., "Ecological Grief as a Response to Environmental Change: A Mental Health Risk or Functional Response," *International Journal of Environmental Research and Public Health* 18, no. 2 (2021): 734–44, doi:10.3390/ijerph18020734.

5. US Department of Health and Human Services. "Asthma and African Americans." (2024): https://minorityhealth.hhs.gov/asthma-and-african-americans

6. Terra Léger-Goodes et al., "Eco-anxiety in Children: A Scoping Review of the Mental Health Impacts of the Awareness of Climate Change. *Frontiers in Psychology* 13 (2022): doi:10.3389/fpsyg.2022.872544

7. Maria Ojala, "Facing Anxiety in Climate Change Education: From Therapeutic Practice to Hopeful Transgressive Learning." *Canadian Journal of Environmental Education*, 21 (2016): 41–56, https://eric.ed.gov/?id=EJ1151866.

8. Chai Tyng et al., "The Influences of Emotion on Learning and Memory," *Frontiers in Psychology* 8 (2017): 1454. doi:10.3389/fpsyg.2017.01454.

9. https://climateawakening.org/

10. Pei Yi Lim et al., "A Guide to Nature Immersion: Psychological and Physiological Benefits." *International Journal of Environmental Research and Public Health* 17 (2020) : doi:10.3390/ijerph17165989

11. Asta Cekaite, "Socializing Emotionally and Morally Appropriate Peer Group Conduct through Classroom Discourse," *Linguistics and Education,* 24, no. 4 (2013): 511–22, doi:10.1016/j.linged.2013.07.001.

12. Maria Ojala, "Hope in the Face of Climate Change: Associations with Environmental Engagement and Student Perceptions of Teacher's Emotional Communication

Style," *The Journal of Environmental Education*, 46, no. 3 (2015): 133–48, doi:10.1080/00958964.2015.1021662.

13. Maria Ojala, "Safe Spaces or a Pedagogy of Discomfort? Senior High-School Teachers' Meta-Emotion Philosophies and Climate Change Education," *The Journal of Environmental Education*, 52, no. 1, (2021): 40–52, doi:10.1080/00958964.2020.1845589.

14. Maria Ojala, "Facing Anxiety in Climate Change Education: From Therapeutic Practice to Hopeful Transgressive Learning," *Canadian Journal of Environmental Education* 21 (2016): 41–56, https://eric.ed.gov/?id=EJ1151866.

15. Maria Ojala," Coping with Climate Change among Adolescents: Implications for Subjective Well-Being and Environmental Engagement," *Sustainability* 5 (2013): 2191–2209, doi:10.3390/su5052191

16. Leslie Davenport, *All the Feelings Under the Sun: How to Deal with Climate Change* (Washington, DC: American Psychological Association, 2021).

17. Maria Ojala and Hans Bengtsson, "Young People's Coping Strategies Concerning Climate Change: Relations to Perceived Communication with Parents and Friends and Proenvironmental Behavior," *Environment and Behavior* 51, no. 8 (2019): 907–35. https://doi.org/10.1177/0013916518763894.

18. Ojala, "Safe Spaces."

19. Meghan Wise, *Climate Doom to Messy Hope: Climate Healing and Resilience*, (Vancouver, BC: UBC Climate Hub, 2022), https://ubcclimatehub.ca/project/climate-doom-to-messy-hope-climate-healing-and-resilience/#:~:text=This%20resource%20is%20for%20everyone,into%20their%20network%20or%20practice.

20. Peter Pellitier et al., "Embracing Climate Emotions to Advance Higher Education," *Nature Climate Change* 13 (2023): 1148–50, doi:10.1038/s41558-023-01838-7.

CHAPTER 15

Developing an EcoJustice Consciousness for Classroom Teachers and their Students

Marissa E. Bellino and Greer Burroughs

"Everything comes into existence via relationships."[1] While this statement seems obvious in so many ways, the sensing and remembering of these relationships—whether social, ecological, or pedagogical—needs to be felt via experiences. Presented in this chapter are the reflections from preservice and in-service teachers who participated in a teacher professional development experience for building EcoJustice consciousness in which relationships to others, to place, and to our teaching was centered. Drawing upon critical place-based pedagogies, we brought together preservice and in-service teachers and created opportunities to collectively engage in sense-making and knowledge production, while deepening commitments and support for teaching with an EcoJustice consciousness. In this chapter, we provide an overview of an EcoJustice professional development weekend for early career educators, share experiences from the preservice and in-service teachers who attended the program, and highlight outcomes from this work. Although some of what we will share are the tangible lessons produced, much is the intangible transformations that can occur when individuals come together for a common goal and commitment to EcoJustice. We end with some lessons learned about doing EcoJustice education at this particular moment.

Conceptualizing EcoJustice Education

An EcoJustice approach in education can address the growing global and local environmental and social crises associated with climate change. We have synthesized relevant literature on EcoJustice education and have conceptualized that educators with an EcoJustice consciousness carry the following attitudes and behaviors into their classroom, curricular, and student relationships:

- An acceptance and valuing of the interdependency of all systems, where individuals see themselves as part of larger systems.[2]
- A recognition of existing hierarchical systems that position humans over nature and some humans as supreme over others, as well as the damage to people and the environment this way of thinking has caused.[3]
- An analysis of cultural, political, and economic forces that create and sustain unjust and unsustainable practices and beliefs.[4]
- A recognition that individuals participate in systems of domination and must uncover the forces that have unknowingly shaped deficit views and participation in damaging practices toward people and the environment.[5]
- An approach to learning that is contextualized and localized; engaging students in researching problems in communities and seeking solutions, while deconstructing the cultural roots of the problems.[6,7]
- An emphasis on working within local systems, forming partnerships, and valuing the knowledge and traditions held by locals as vital in the problem-solving process.[8]

An EcoJustice framework is the theoretical lens that shapes our work, reimagining teacher preparation, professional development, and the creation of classroom resources. Martusewicz, states, "This field that we call EcoJustice is not 'just about the environment.' Our first task is to analyze the cultural roots of all sorts of intersecting social and ecological violence that are created when we naturalize the superiority of some humans over others, and all humans over other living creatures."[9] In an effort to analyze these cultural roots, we argue that "education must help students develop the skills and habits to critique the cultural norms, structures, and forces at work in society that operate to constitute and reproduce unjust and unsustainable attitudes about other people, other living beings, and the land."[10]

We draw on "critical pedagogies of place"[11] to investigate complex socio-environmental issues. As Gruenewald explains, this "focus on the lived experience of place puts culture in context, demonstrates the interconnection of culture and environment, and provides a locally relevant pathway for multidisciplinary inquiry and democratic participation."[12] By explicitly learning about and questioning the multiplicity of perspectives individuals and communities have regarding place, deeper critical analysis that links environmental conditions to social structures emerges.

Education for sustainability (EfS) also informs our work by offering guidelines that call for developing knowledge, respecting the commons, and taking responsible individual and collective action.[13] We are also guided by the standards to support climate change learning in New Jersey, which call for interdisciplinary education that includes "authentic learning experiences that integrate a range of perspectives and are action oriented."[14] Finally, our work draws on participatory research methodologies that include multiple opportunities for and forms of reflection and maximize participants' meaning-making do themselves. Raising one's awareness of socio-environmental issues and developing a critical consciousness are important precursors in developing and delivering Climate Change Education and EfS lessons from an EcoJustice perspective. A critical consciousness can be achieved through acquisition of knowledge, coupled with opportunities for individual meaning-making.

EcoJustice Professional Development

Critical place-based learning through an EcoJustice lens that is rooted in relationships between communities and educators was at the heart of an EcoJustice professional development held in the spring of 2023. A group of twelve female-identified preservice (eight) and in-service (four) teachers met for a three-day professional development weekend in the mid-Atlantic region. The goals were to nurture a community of practice among participants around EcoJustice education, provide critical place-based learning experiences, and create tangible multidisciplinary instructional materials. Two critical place-based experiences grounded the weekend in the larger themes of climate change, water, and land, including a tour of a local watershed and a visit to a local permaculture farm and farmers' market. Additional activities included eating at a farm-to-table restaurant, participating in a workshop on hydroponics, and viewing a film, *Rotten, A Sweet Deal*,[15] on sugar production in the United States. The learning experiences also incorporated elements of mindfulness meditation; honoring Indigenous and traditional ecological knowledge; and time for reflection, journaling, critical dialogue, and lesson collaboration. Teams of in-service and preservice teachers worked together at the K–3, 4–6, and 7–8 grade levels to create educational materials that captured their takeaways from the weekend. Lessons created focused on food access, production, and consumption; watershed health and protection; and the health of soil ecosystems as impacted by local land-use decisions.

One goal of the weekend was to nurture a desire among participants to find value in their work and be part of a solution. All the attendees were entering or had recently entered a field rife with controversy and at a time of grave upheaval in education due to COVID-19. Additionally, they live in a nation that is increasingly politically divided and are exposed to a constant barrage of news about war, natural disasters, and impending climate change threats.

Current times are trying for many, but educators are on the front lines of attacks from parent organizations calling for books to be banned, policymakers drafting legislation that impacts how teachers can discuss race and other "controversial issues," and constant pressure to raise test scores and make up for academic declines brought on by distance learning during COVID-19. In this context, large numbers of teachers are leaving the field; a 2022 Gallup poll[16] found that teachers suffer the highest rate of burnout among all professions. With this backdrop, the following question took high priority during the weekend planning and execution: How can we not only provide resources and knowledge for participants to address issues of EcoJustice, but also provide them with the social support to remain committed to their convictions and the connections formed over the weekend? We realized that in many ways, we needed to feed the minds, bodies, and hearts of our participants *during* and *after* the professional development.

BUILDING CONNECTIONS

Fostering connections to the earth, our personal beliefs, and communities doesn't necessarily come naturally. Therefore, we provided intentional opportunities to facilitate these connections. This began with discussing practices of Native People, specifically the Lenape, who had once inhabited the land we were on. Curtis Zunigha, a descendant of the Lenape and member of the Delaware Tribe of Indians, explained the connection to the earth in the following way: "Lenapehoking [the land of the Lenape] is the land, it's the waters, the rivers, the lakes, even the ocean, all of the cosmos, all of our connection with the earth, the waters, the sky, the animals, all life, the mountains, the ancient ones, the ones that have the ancient memory, all of these things, to the Lenape have a spirit."[17]

For some participants, it was moving to recognize that the land we were on once belonged to the Lenape. We talked about what we could learn from the Lenape and why their history wasn't better taught in our school programs. Then, we presented "Prayer to Mother Earth" as a cultural artifact that provided insight into the reverence and respect for earth among many native peoples.

PRAYER TO MOTHER EARTH[18]

We return thanks to our mother,
the earth, which sustains us.
We return thanks to the rivers and streams,
which supply us with water.
We return thanks to all herbs,
which furnish medicines
for the cure of our diseases.
We return thanks to the corn,

and to her sisters, the beans and squash,
which give us life.
We return thanks to the bushes and trees,
which provide us with fruit.
We return thanks to the wind,
which, moving the air,
has banished diseases.
We return thanks to the moon and the stars,
which have given us their light
when the sun was gone.
We return thanks to our grandfather He-no,
who has given to us his rain.
We return thanks to the sun,
that he has looked upon the earth
with a beneficent eye.
Lastly, we return thanks to the Great Spirit.
in whom is embodied all goodness.
and who directs all things,
for the good of his children.

The group expressed such a strong affinity for the sentiments and knowledge embedded in the words that we all agreed to read the message of the prayer each morning to center our work.

In addition to reading the Iroquois prayer, participants were invited to voluntarily join in a guided meditation focused on grounding and connecting to the earth. While some of these methods are not typically present at educational professional development programs, we believed these aspects helped participants connect to themselves, the earth, and cultural practices that honor nature and foster mindfulness. Reflections from many of the teachers indicate we met these goals, as will be seen in their reflections from the weekend.

REFLECTION FROM PARTICIPANTS ON ECOJUSTICE PROFESSIONAL DEVELOPMENT

At the end of the weekend, we asked participants to reflect upon the following questions and add additional questions they were curious about.

- Why do you think EcoJustice is a powerful lens for teaching in your context?
- How did your EcoJustice consciousness originate and how does it continue to evolve?
- How can those of us in teacher preparation continue to support you and others in efforts to infuse EcoJustice consciousness into classroom practice?

The following additional student questions were included:

- Do you find it challenging to infuse experiences, practices, and knowledge from the professional development? If so, what?
- How have you been able to fund or gain support in your school community to further teach ecojustice education?
- Do you find it challenging to modify ecojustice education for younger learners (K–2)?
- When you feel like you are losing the spark/passion you had during the weekend itself in the normal day-to-day classroom, what have you done to regain it?

Themes of social, ecological, and pedagogical connections ran deep in participants' reflections. One teacher remarked,

> I am excited to feel connected to a community of educators that this work matters to. I think this weekend will be beneficial for myself. Over this first year of teaching, I have not made much time truly for myself or without thinking about what is coming next. I am hopeful I can be present during this weekend. Not being present has made me feel disconnected to myself this school year. (Alea, middle-school English language arts teacher).

It is often the case that new teachers struggle to pause for reflection and that a large part of the value for the group was the sense of community, reflection, and presence that was intentionally cultivated.

Additionally, connection to and disconnection from the earth, especially in the geographic context where our participants reside and teach (a densely populated mid-Atlantic state), was a common theme. One preservice teacher commented on what they valued about the weekend, stating, "Connecting myself to the earth and fostering reverence for the earth and my bodily and spiritual connection to it; recognizing everything that nature offers; grounding practices, food, energy, medicine, herbs, teas, etc" (Emma, preservice early childhood teacher). Many others shared Emma's sentiment as we continued to ask questions about ways schools can be sites for cultivating connection to the earth, how it feels when we sense in our beings our connection to the earth, and the rhetoric we encounter every day that severs this connection.

New ways of thinking about classroom practices emerged for all participants, in particular ways of engaging with local resources and experts in explorations of local places and the multidisciplinary nature of socio-ecological issues. Meeting a local permaculture farmer and other local farmers at the farmers' market helped participants value knowledge outside the formal curriculum found in schools.

One participant commented, "Something I learned this weekend was how to incorporate local people into learning and having nontraditional educators come into the classroom to teach students" (Hannah, preservice elementary teacher). Jennifer, a secondary biology preservice teacher noted,

> What I learned is the importance of how everything connects rather than just knowing the concept in a more biological sense . . . it was very useful that I gained exposure to learning about the watershed and the relationship between the water, soil, and the animals and simply how if we aren't taking care of one thing, whatever is connected to it will also be impacted.

Both Hannah and Jennifer are now thinking more about expanding classroom teaching and student learning beyond the conventional classroom by connecting to local resources and deeper, more relational ways of connecting science content.

Below, we present highlights from participant narratives that speak to the personal and professional impact of the professional development weekend.

Alex, preservice early childhood teacher. During the weekend, I felt an immense sense of privilege to be surrounded by a room full of intelligent females, to be able to hear from educators who were already in the education field, and to recognize the power we truly have as educators to make a change. Going into the PD, my mind-set around teaching was framed from the curriculum we were always told we had to follow. Hearing the in-service teachers talk about how they were able to successfully implement EcoJustice into their curriculum made me hopeful for our future generations and eager to learn more about how I could do the same. My team focused on implementing EcoJustice into early childhood education lessons and we honed in on the farm-to-table experiences we saw firsthand. Knowing that we wanted to incorporate these ideas into our lesson plans and unit, we focused on answering the following questions: Where does food come from? And essentially, how does it get to us?

Emma, preservice elementary teacher. On the first day, with my bare feet touching Mother Earth, connected in a circle of empowering, powerful women, I felt the value of EcoJustice education. It was at this moment that I understood the importance of educating toward diverse, sustainable, and democratic societies. Creating, nurturing, and justifying reverence for the earth and its deep interconnectedness to every and all social justice and global systems is my understanding of EcoJustice education. Nature has an inherent connection to social justice issues that is often ignored, neglected, and even denied in schooling. Over the course of the weekend, I found myself having many conversations that challenged my thinking that had been previously ingrained into my mind as "normal" or "correct." What helped me grow the most as an educator were the meditation sessions that allowed us to intentionally, consciously, and physically connect to the ground beneath our feet. I realized it would forever be my job to protect Mother Earth and teach others how to conscientiously do the same.

Morgan, in-service middle school math teacher. I have had the privilege of working on the research Ecojustice team since 2017. I continually gravitate toward these experiences because of the community that is cultivated. The fresh perspectives and intellect of young preservice educators, in combination with the experiences and practical application of in-service teachers creates a powerful experience for all. A barrier that I have faced is the implementation of EcoJustice lessons. Because the school that I work in utilizes a structured curriculum, I was concerned that incorporating anything we created would be pushed back on, "extra work," or something I saved for the end of the year when I had "nothing else to do." However, I found I was able to use the mindset cultivated from the weekend in everyday lessons with no change in the curriculum I was already using. Upon returning to the city that I work in, I was also viewing it through more of a green lens and am more aware of parks and green areas around my school. As educators, it is our obligation to identify the importance of spending time in green spaces and create opportunities for our students to learn in and about these spaces.

Paige, in-service third-grade teacher. "The ground begins beneath our feet." This powerful quote stood out early on during the weekend. Being an educator, it can sometimes seem overwhelming to teach EcoJustice education due to the number of issues and topics one could cover. However, this quote and PD made me excited to discover ways to bring EcoJustice issues into my everyday teaching. The community of strong, intelligent women educators that came together to discuss their passions toward EcoJustice education was inspiring. I was fortunate to learn from past professors, local experts, and preservice teachers about topics regarding environmental and social justice issues that our community is facing today. Using our newly gained knowledge, we created unit lessons focused on water usage and advocacy for grades 3 to 5. What was so amazing was that we were able to do this with many topics that we had learned that weekend. As a group, being able to take a big issue, collaborate passionately with one another, and turn it into an amazing unit for students where they can develop solutions using a place-based learning experience was a wonderful full-circle moment that is often missed in the everyday classroom.

Reflections for both preservice and in-service teachers illustrate the range of impacts the weekend had on participants. In the next section, we present the experience of implementing lessons for two participants, Alex and Emma. Each translated their emerging EcoJustice consciousness into their first-year teaching in third grade and pre-K, respectively.

Translating Professional Development into Classroom Teaching

Alex and Emma elected to work with us on implementing lessons inspired by the weekend in the fall of 2023. Both were new teachers and our former students. Each faced challenges in the endeavor, but with some support from us and each other, they found places in their teaching where they could bring aspects of EcoJustice to their students. We met via Zoom three times to plan and reflect. The initial meeting focused on discussion of upcoming topics to be taught, requirements of their curriculum, and factors such as expectations from parents and colleagues. The meeting offered a space for Alex and Emma to navigate these topics and brainstorm and flesh out ideas. We offered guidance with the state standards and our knowledge of science, social studies, and EcoJustice, but each made their own decisions based on the contexts in which they were teaching. Summaries of their journey are presented next.

ALEX

Alex was excited to start her first-year teaching in a third-grade classroom. Her new job was in an upper-middle-class suburban community where high academic standards and outcomes were the norm. According to the online resource Niche, which reviews data and rates P-12 schools, her school was "a top-rated public school district" with an A+ rating.[19] She knew she had to maintain the high standards expected by parents, administrators, and her colleagues. This was not in itself a problem, as Alex was a bright, high-achieving, and motivated individual. The issue for Alex was more about how she could integrate topics that were important to her into her teaching and still follow the existing curriculum.

Attending the EcoJustice professional development weekend seemed to provide Alex with guidance for how she could achieve her goals. She wrote in her reflection from the weekend, "My mindset around teaching was framed from the curriculum we were always told we had to follow. Hearing the in-service teachers talk about how they were able to successfully implement EcoJustice into their curriculum made me hopeful for our future generation and eager to learn more about how I could do the same." She found learning from practicing teachers who were graduates from the same teacher education program as she was inspirational. The opportunity to collaborate with others that weekend on creating EcoJustice lessons for early childhood classrooms proved to be instructional. With that experience in mind and support from us and Emma, Alex felt ready to find ways through the existing curriculum.

When we began the work, Alex was very conscious of her role as a new teacher. She wrote, "One concern I had about going beyond the curriculum was the risk factor. Naturally, I was worried about how it might be perceived by my colleagues, especially those on my third-grade team who are veteran teachers." However, Alex recognized that topics covered in the third-grade science curriculum on water and climate could serve as a basis for further exploration with her students. Even though this was a natural connection, she still expressed concerns of going "too far off the track" during the initial stages of planning her unit. An important moment in our early conversations with Alex was when we discussed whether she had freedom to meet the curriculum goals as she saw fit or if she had to follow the plan of the rest of the third grade. She admitted she probably had more freedom than she had been willing to exercise and once she recognized this, the ideas flowed.

She decided to build on the anchor phenomenon from the school science curriculum, "After a rain, why is there water in some places on the playground and not others?" The curriculum focus question was, "What happens when water falls on different surfaces?" The curriculum content could be naturally connected to larger issues of water use, water management, and climate change. A UN report on water and climate change states, "Climate change is primarily a water crisis. [...] However, water can fight climate change. Sustainable water management is central to building the resilience of societies and ecosystems and to reducing carbon emissions. Everyone has a role to play—actions at the individual and household levels are vital."[20] We also drew on information from the Environmental Protection Agency:

> As we develop our cities and towns, we replace forests and meadows with buildings and pavement. And now when it rains, the water runs off roofs and driveways into the street. Runoff picks up fertilizer, oil, pesticides, dirt, bacteria and other pollutants as it makes its way through storm drains and ditches - untreated - to our streams, rivers, lakes and the ocean.[21]

Thus, Alex saw a space in the curriculum to focus the students on issues of water in their surroundings and to make connections to the environment and human interactions. Alex chose to use a photovoice project to help students make these connections. Photovoice involves using images to create opportunities for children and youth to share stories, make connections across experiences, and collectively co-construct knowledge.[22,23] The use of images allows for flexibility in implementation where the methods can be adapted and modified for various age levels. Alex engaged her third-graders in a series of photovoice activities where they took pictures of water and water usage as they saw them in their daily lives and community. Once the students submitted their photos, she engaged them in an analysis drawing on some of the following photovoice questions:

- What do you **see** here?
- Why did you take this picture?
- What is really **happening** here?
- How does this relate to **our** lives?
- Why does this **condition** exist?
- What can we **do** about it?
- Where is the **social studies** in this picture? How can we connect our learning about our community and relationships to this picture?
- Where is the **science** in this picture? How can we connect our learning about water and climate to this picture?
- How could this image **educate** others?
- What do some of our pictures have in common?
- Are there any images or issues that are really different that stand out?
- What further questions can we ask about this picture?

The students then chose two to three of their pictures to write about using some of the following sentence starters:

- This is a picture of . . .
- I took this picture because . . .
- What I see happening in this picture is . . .
- This relates to my life because . . .
- This is happening because . . .
- We can address this issue by . . .
- This picture connects to things we have been learning about in school like . . .
- We can learn _____ from this picture.
- In our groups we all had similar pictures about . . .
- Some of our pictures were really different, like . . .
- This picture makes me wonder about . . .
- If I were to do further research about this situation, I would want to know . . .

The students eagerly engaged in the project, excited at the chance to document how they saw water and water usage in their lives. Table 15.1 shows a sampling of student pictures and reflections. These young students were beginning to make connections between textbook and curriculum concepts to the natural world around them.

Alex realized a next step in the process would be to help the students consider who made the roads or planted the grass and how it related to water runoff or absorption. She recognized that most third-graders need prompting from the

Table 15.1 Examples of Student Photovoice Projects on Water in their Community

There are puddles in some places and not in others. (3rd grade student, November 2023)	I took this photo because it was wet grass…this picture shows that alive grass repels and also that dead grass absorbs. (3rd grade student, November 2023)	This was a morning me and my mom were walking to school and saw this rainwater in the road. (3rd grade student, November 2023)

teacher to ask those questions. However, one student did make larger connections on their own when they wrote about a picture of leaves blocking a sewage drain and they questioned how the leaves got there, how this was a problem for drainage, and how humans could "fix" this problem (see figure 15.1).

Alex was pleased with the outcome of the project, explaining, "Photovoice proved to be a great tool, allowing my students to become more engaged when exploring the connection between water and climate. […] The photos acted as visual aids for students to better view the interactions between humans and the natural environment." Though she did see that many third-grade students need support in making these connections, she still expressed, "The hands-on learning that photovoice provides sparked every student's curiosity and encouraged many 'surface-level' thinkers to think more critically about the impact of the human race on the natural environment." Perhaps most importantly, Alex saw she didn't have to stray too far from the curriculum or her colleagues' practices to integrate a deeper analysis of water usage and the human element and to integrate place-based learning that would help children connect more to their environment. She now feels confident she can repeat and build on this process to help make the connections clearer for students.

EMMA

Emma is a pre-K teacher at a private pre-K with seventeen 3.5–4.5-year-old students and one teacher assistant in her classroom. Students attend the pre-K from

DEVELOPING AN ECOJUSTICE CONSCIOUSNESS 205

Figure 15.1 Student Photo of Sewage Drain Covered in Leaves from photo-voice project (3rd grade student, November 2023)

8:30am to 5:30pm five days a week, with two explicit times for instruction each day: once in the morning and once in the afternoon. The pre-K curriculum offered Emma flexibility in choosing a theme each month to teach, allowing more cross-curricular opportunities for learning with her students. She also has access to outdoor spaces that she and her class are allowed to utilize daily. Without the constraints of a strongly mandated curriculum, Emma decided to express her emerging EcoJustice consciousness through various pedagogical strategies, including photo-elicitation, read-alouds, crafts, play, and whole-group discussion. She drew on diverse teaching strategies in emergent ways, responding to her students as they were making meaning and connections to prior knowledge and collectively generating new knowledge.

Working with early childhood students to create developmentally appropriate lessons was the most challenging aspect for Emma when it came to engaging her students in EcoJustice-oriented lessons. However, this didn't stop her from trying to infuse aspects meaningful to her from her time at the EcoJustice weekend into her classroom, including empathy, justice, and ecologically oriented themes. Emma utilized photo-elicitation and read-alouds to prompt her students' feelings and emotions, as well as to generate questions and inquiries that went on to prompt further learning. Some of the images she used highlighted inequality, equality, equity, and justice from *The Giving Tree*, close-up images of leaves and root systems, and pictures of human hands. These images speak to Emma's deep commitment to social justice, as well as how she made space for students to make connections between the function of veins in leaves, the roots of plants, and the human hand. Children's books she read with her class included *A Garden to Save the Birds, Bird Builds a Nest, The Proudest Blue, The World Needs More Purple People*, and *I Am Human: A Book of Empathy*.

Emma shared that her goals for her students were to create a safe environment for inquiry, discovery, and questions; to help students make connections between the natural and social environment; and to see the beauty in biological and human diversity. Reflecting further on how she hopes to continue to engage the students in her pre-K classroom in EcoJustice education, she responded:

> Ecojustice education has become a deeply rooted part of my teaching philosophy that inspires me to create change and create and nurture environments in which students feel safe and comfortable advocating for change and learning the tools necessary to help facilitate them as responsible democratic learners. I believe as a teacher it is not only my responsibility, but my duty to each and every student to prepare them to be facilitators of social change and advocate for the needs of themselves and the communities around them. It is my hope that I will be able to continue to engage my students in connecting with environmentally friendly and social justice practices to create a reverence for the earth, as well as an intersectional lens through which to view the world.

Lessons Learned about Nurturing and Implementing EcoJustice Education

The work of engaging preservice and in-service teachers in EcoJustice education efforts continues to offer many lessons for how to do this within the constraints and realities of education today. The following is a set of interconnected lessons we learned through our EcoJustice professional development about how to support teachers in forging forward with bringing an EcoJustice consciousness into the classroom.

- *Be intentional as faculty in Schools of Education in supporting preservice teachers in gaining the skills and expertise to bring socio-environmental issues into their future classrooms.* This can occur in science and social studies methods courses, as well as through professional development opportunities.
- *Work with alumni as they transition into full-time teaching spaces to continue to feel connected to other teachers who share a commitment to EcoJustice.* This can be done through professional development opportunities, as well as cultivating social networking and resource-sharing spaces.
- *Work with in-service teachers to navigate the curriculum and mine spaces for EcoJustice lessons within the current state standards and district curriculum.*

- *Engaging educators in critical place-based learning allows participants to feel for themselves the power of this pedagogy, which in turn can inspire them to enact it in their classrooms.* Allowing time and space to collectively reflect upon the value for teaching, learning, and relationship-building that critical place-based learning affords all participants helps illuminate ways similar experiences can be transferred to their teaching contexts.
- *Support new teachers in finding local resources for their respective communities and tap into existing resources (i.e., state Departments of Environmental Protection, environmental and social justice community organizations, local historical societies, etc.).*
- *Help teachers feel more connected to themselves, a community of teachers, and their environment.* This can be done through intentional practices such as mindfulness, journaling, group reflections, and collaborative planning.

Conclusions

Our disciplinary and standards-based curricular culture has severed the natural relationship between science and social studies, limiting teachers' flexibility, creativity, and time dedicated to learning about socio-environmental issues. Meanwhile, climate change, inequitable access to sustainably grown and nutrient-dense foods, and the declining health of watersheds that provide our drinking water are all realities children and youth growing up today will be forced to reckon with in their lifetimes. If we have any hope of nurturing and educating a future society that is equipped with responding to these interconnected crises, we need to begin by reconnecting our bodies, minds, and hearts to the earth and our communities. We have seen the power of engaging educators in critical place-based pedagogies that foster an EcoJustice consciousness and the ways these experiences can heal these severed connections. Our future teachers' commitment to a just and sustainable future must be continually nourished so they can provide opportunities for their students to make deep connections to one another and the earth and heal our collective separateness.

NotesNotes

1. Rebecca A. Martusewicz, "EcoJustice for Teacher Education Policy and Practice: The Way of Love," *Issues in Teacher Education* 27, no. 2 (2018): 17–35, https://www.itejournal.org/wp-content/pdfs-issues/summer-2018/05martusewicz.pdf

2. Joseph A. Henderson, "Out of Sight, Out of Mind: Global Connection, Environmental Discourse and the Emerging Field of Sustainability Education," *Cultural Studies of Science Education* 10 (2015): 593-601, doi:10.1007/s11422-014-9641-z.

3. Martusewicz, "EcoJustice for Teacher Education."

4. Rita Turner and Ryan Donnelly, "Case Studies in Critical Ecoliteracy: A Curriculum for Analyzing the Social Foundations of Environmental Problems," *Educational Studies* 49, no. 5 (2013): 387–408, doi:10.1080/00131946.2013.825262.

5. John Lupinacci and Alison Happel-Parkins, "Ecocritical Foundations: Toward Social Justice and Sustainability," in *The Social and Cultural Foundations of Education: A Reader,* ed. Joshua Diem (Solana Beach, CA: Cognella Academic Publishing, 2016), 34–56.

6. Ethan Lowenstein, Rebecca Martusewicz, and Lisa Voelker, "Developing Teachers' Capacity for Ecojustice Education and Community-Based Learning," *Teacher Education Quarterly* 37, no. 4 (2010): 99–118, https://www.jstor.org/stable/23479462

7. Lupinacci and Happel-Parkins, "Ecocritical Foundations."

8. Chet A. Bowers and Rebecca Martusewicz, "Ecojustice and Social Justice," In *Encyclopedia of the Social and Cultural Foundations of Education,* ed. Eugene F. Provenzo, Jr. (Thousand Oaks, CA: SAGE, 2009), 272–79.

9. Martusewicz, "EcoJustice for Teacher Education," 20.

10. Turner and Donnelly, "Case studies," 388.

11. David A. Gruenewald, "The Best of Both Worlds: A Critical Pedagogy of Place," *Educational Researcher* 32, no. 4 (2003): 3–12, doi:10.3102/0013189X032004003.

12. David A. Gruenewald, "Place-Based Education: Grounding Culturally Responsive Teaching in Geographical Diversity," in *Place-Based Education in the Global Age,* ed. David A Gruenewald and Gregory A. Smith (New York City: Routledge, 2014), 137–54.

13. Bethany Vosburg-Bluem, Margaret Crocco, and Jeff Passe, eds., *Teaching Environmental Issues in Social Studies: Education for Civic Sustainability in the 21st Century* (Silver Spring, MD: National Council for the Social Studies, 2022).

14. "Climate Change Education by Grade Band," New Jersey Department of Education, January 1, 2024, https://www.nj.gov/education/climate/learning/gradeband/.

15. *Rotten*, season 2, episode 4, "A Sweet Deal," directed by Lucy Kennedy, aired October 4, 2019, Netflix.

16. Devlin Peck,. "Teacher Burnout Statistics: Why Teachers Quit in 2024," January 11, 2024, downloaded at https://www.devlinpeck.com/content/teacher-burnout-statistics.

17. "Plants as Medicine: Lenape Healing Traditions Continue Today," Columbia School of Nursing, April 20, 2021, https://www.nursing.columbia.edu/news/plants-medicine-lenape-healing-traditions-continue-today

18. Mark Linden O'Meara, "Prayer to Mother Earth: An Iroquois Prayer," Spirituality Practice, https://www.spiritualityandpractice.com/practices/practices/view/27383/prayer-to-mother-earth

19. "Westfield, NJ," Niche, January 2, 2024, https://www.niche.com/places-to-live/westfield-union-nj/.

20. "Water and Climate Change," United Nations, https://www.unwater.org/water-facts/water-and-climate-change.

21. "Soak Up the Rain: What's the Problem?" U.S. Environmental Protection Agency, November 8, 2023, https://www.epa.gov/soakuptherain/soak-rain-whats-problem.

22. Caroline Wang and Mary Ann Burris, "Photovoice: Concept, Methodology, and Use for Participatory Needs Assessment," *Health Education & Behavior* 24, no. 3 (1997): 369–87, doi:10.1177-109019819702400309.

23. Marissa E. Bellino, "Using Photovoice as a Critical Youth Participatory Method in Environmental Education Research," in *Doing Educational Research: A Handbook* (Boston: Brill Publishing, 2015), 367–82.

CHAPTER 16

High-Quality Instruction Begins with Great Resources

Beverly Plein, Julia T. Sims, and Margaret Wang

A recent Google search of climate change education resulted in more than 1.5 trillion hits from a wide range of sources, including the United Nations, universities, NASA, government agencies, advocacy groups, policy centers, and news publications. According to a national survey of US educators, finding resources that lead to high-quality instruction is a difficult task.[1] Research suggests that a lack of access to high-quality resources has resulted in students engaging in activities that are below grade level, and this is particularly true in schools with a large number of students of color, those from low-income families, and/or English-language learners.[2] In this chapter, the term *resources* means anything that aids in the instruction or assessment of student learning (e.g., books, articles, databases, videos, photographs, podcasts, websites, individuals with relevant expertise or experiences, simulations, etc.).

The first section of this chapter will discuss resources designed to support high-quality climate change learning experiences in formal and informal educational settings. Generally, high-quality resources are culturally responsive and inclusive, accurate and standards-based, and support student-centered and action-oriented learning environments. Given the ever-changing nature of resources, this section will describe these resources' characteristics in a generic manner rather than naming specific resources. The second section of this chapter will include vignettes including students in different grade levels to illustrate the recommendations described in the first section. At the end of the chapter, there are questions to guide the selection of resources. Readers are encouraged to adapt the questions to meet the needs of their specific learning environments.

Culturally Responsive and Inclusive Resources

At the core of providing high-quality instruction for all learners is the commitment to enact culturally responsive and inclusive approaches to learning. These two frameworks are closely related because they promote instruction that considers the individual's learning needs and experiences. Culturally responsive teaching specifically takes into consideration the impact a student's background and experiences can have on their learning. An inclusive learning environment or one in which the teacher uses a universal design for learning approach recognizes the need to identify and remove barriers to learning.

As a first step in evaluating a resource's usefulness, educators might ask themselves how a resource will contribute to a unit that values different identities and perspectives (e.g., racial, cultural, religious, gender, sexual orientation, age, political). For example, does the resource support students whose preparation for learning might have been impacted by their socioeconomic status or language ability? Relatedly, does the resource contribute to a learning environment that acknowledges there are many paths to understanding content and multiple ways to demonstrate learning? In addition, does the resource contribute to a unit that provides multiple entry points to student interests and ability levels? Does it reflect the various cultural identities of the students and their communities? Is the information provided available in different formats, languages, and reading levels? For example, educators may want to seek out websites that provide resources available as videos and podcasts in different languages, as well as articles written at different reading levels and in different languages to remove barriers for students who are not reading on grade level. Resources that promote inclusivity and aid in differentiated instruction will likely lead to increased student engagement and achievement.[3]

Accurate and Standards-Based Resources

The internet provides unlimited access to information; however, we should not assume every resource provides accurate and relevant information. It is critical that educators and students have the tools and take the time to evaluate a resource's credibility and usefulness before selecting it for use. For example, before using a resource, it is important that educators and students understand the context associated with its creation to better understand its value. Asking questions about its purpose, when it was created, and the author's expertise will help determine the context of a particular resource. Specifically, educators should evaluate whether and what biases might be present based on the resource's origins. It is worth noting that while resources that exhibit bias should typically be avoided, they can be beneficial in illuminating specific perspectives associated with particular time periods, individuals, and organizations and may serve a purpose in some instructional units.

Moreover, effective instructional experiences incorporate an integrated approach to acquiring content knowledge, skills, and practices necessary for college and career readiness.[4] In addition to a wide variety of educational institutions, many agencies and organizations have identified the knowledge and skills students will need to be college- and career-ready. For example, student learning standards have been developed by state departments of education and national professional organizations (e.g., Next Generation Science Standards [NGSS]) to describe these expectations by grade or grade-band levels. Understanding the standards' expectations for specific grades, as well as earlier and later grades, is essential in determining whether a resource can support high-quality instruction for the targeted students. In reviewing a resource for its usefulness, an educator may want to ask if the resource will contribute to students learning knowledge and skills associated with their grade. Likewise, it is important to consider whether *all* students have the necessary foundational knowledge to use the resource and if scaffolding can be provided to aid in their understanding.

It is worth noting that leveraging the expectations described in student learning standards as part of the resource evaluation process should promote coherence across grades. In a recent survey, US educators indicated a lack of access to resources that provided a coherent approach to learning.[5] This is problematic because foundational knowledge is often needed to understand more complex ideas related to climate change.[6] Student learning standards can also be useful for individuals working in nonformal learning settings to better understand the background knowledge students may have. Standards may also aid informal educators in selecting useful resources that may *not* be found in school settings.

Regardless of students' ages or the setting where the learning will take place, resource selection should go beyond addressing content knowledge aligned with science standards. Given the social, political, and economic impacts associated with climate change, an integrated and comprehensive approach to learning about climate change across content areas (e.g., mathematics, social studies, literacy, health, visual arts) is necessary. In addition to learning about factors that contribute to climate change and its impacts, students will also need to develop learning and innovation skills (e.g., critical thinking); media skills (e.g., technology); and life skills (e.g., decision-making).[7] To evaluate a resource based on these criteria, educators may ask themselves, *Will this resource contribute to students' understanding of the impacts of climate change across content areas? Will the resource aid students in developing essential skills for college and careers?*

Resources for Student-Centered Learning and Action-Oriented Environments

Student-centered learning environments reflect the types of experiences that happen in the *real world* and are designed to be *relevant* and *meaningful* to students. This contrasts with more traditional instructional approaches that limit resource use to textbooks and videos, where students consume information by *reading* about or *watching* videos and demonstrate their understanding only to their teacher. In student-centered learning spaces, educators are likely to leverage a wide variety of resources, including those that enable students to engage in real-world processes using authentic tools, help students address an issue in the school or community, or that allow students to demonstrate their learning to an authentic audience.

For example, an observer of a student-centered learning space might see students engaging with resources that enable them to access data stored in a government-sponsored database about the topic under study, collaborate in groups to discuss their findings, and then share their claims with peers to receive feedback. In this type of student-centered learning environment, students benefit from using "real-world" resources such as online databases created with geographic information system tools and simulations that provide data about energy consumption, atmospheric conditions, etc. In addition, digital tools that support collaboration and information-sharing provide opportunities for students to receive feedback asynchronously or synchronously (e.g., through web-based documents). Before engaging students in this type of real-world scenario, it is important to consider if they have sufficient foundational knowledge to effectively use the resource. Similarly, it is important to consider whether students have the skills to effectively use the digital resources being provided and if they are age-appropriate (especially if the resource allows students to interact with others outside the learning setting).

Additionally, educators can leverage climate change issues happening in the community or a nearby area to enhance relevance and promote learning. Research suggests that local and state experts and agencies can serve as valuable resources because of their experience and expertise with issues in the students' community.[8] Educators of young children may want to vet these individuals and organizations before inviting them to participate, whereas older students might benefit from identifying individuals with relevant expertise and connecting directly with them. For example, the state climatologist, a professor engaging in relevant research, or an individual representing an organization that is actively involved in a local or statewide climate change issue could potentially provide information not typically found in more traditional reading materials. Further, these types of interactions may provide the opportunity for students to develop other types of skills (beyond content knowledge) needed in the real world. For example, preparing for and engaging with an individual with expertise may help students develop important speaking and listening skills. Moreover, meeting professionals in different fields exposes students to potential career possibilities. Before interacting with individuals outside the learning setting, it is important to consider what expertise the individual has, the background knowledge students will need to understand the information that will be shared, and in what ways the information might be biased.

Climate change education research suggests that engaging students in action-oriented learning is important, so they do not feel a sense of despair.[9] Often known as project- or problem-based, experimental, inquiry-based, or student-centered learning, these types of experiences provide opportunities for students to create products or engage in projects using real-world tools for an authentic audience. Regarding climate change education, action-oriented learning refers to those types of learning experiences that provide opportunities for students to work toward solutions or engage in actions that address climate change impacts. This might include studying a local issue and sharing information or possible solutions with community members or government officials.

Action-oriented learning can occur when students participate in citizen science projects; state, national, or global climate challenges (e.g., NatGeo's Slingshot Challenge, Lexus Eco Challenge); or student advocacy opportunities organized by educational institutions, government agencies, nonprofits, businesses, or industries interested in addressing climate change impacts. For example, a citizen science project like NASA's Chesapeake Bay Watch enables students to engage in data collection using authentic equipment and smartphone applications to aid in monitoring coastal waters. Regardless of who sponsors the activity, of key importance is the use of resources for this type of learning experience. Action-oriented learning environments move beyond spaces where students complete worksheets, watch videos, or take tests to demonstrate their learning. Instead, students engage in authentic learning experiences by observing and documenting what is happening in the real world and demonstrate their learning and proposed actions to audiences in an authentic way. An observer of an action-oriented learning environment will see students selecting and using resources in a similar manner to professionals in the field.

As the implementation of climate change education becomes more prevalent, it is essential to understand how to select resources likely to support high-quality instructional experiences.

It is worth noting that resources alone will not necessarily result in effective instruction. (For more information about high-quality climate change instruction, see Chapter 13: Sustainability and Climate Change for All, A Systematic Approach to Implement K–12 Problem Based Learning and Design Thinking, and Chapter 20: Effective Climate Change Education in the Classroom Starts with Effective Climate Change Professional Learning for Educators, for information about high-quality professional development.) In the first section of this chapter, a description of the importance of selecting culturally responsive, inclusive, accurate, and standards-based resources was provided. Then, the types of resources commonly found in student-centered and action-oriented learning spaces were discussed. In the next section, several vignettes are provided to illustrate how resources with the characteristics described earlier are used to facilitate learning about climate change.

Vignette 1: Climate Project

A New Jersey teacher leveraged Big History's free Climate Project curricular resources to develop a unit that reflected the state's student learning standards while promoting an action-oriented approach to local climate change issues. Juniors and seniors in this elective course had access to a wide variety of web-based resources, including articles written at different reading levels, audio recordings, and videos (with transcripts), all of which had been vetted by scholars and reviewed by educators. The teacher explained that students whose first language was not English specifically benefited from the transcripts associated with the videos because they allowed the students to work at their own pace as they became familiar with new, academically challenging words.

Students collaborated in small groups and completed a variety of activities to better understand how climate change has impacted individuals and countries. For example, the teacher, serving in the role of "lead learner," guided the students in the use of an online simulation (i.e., En-Roads) to answer questions they had about how climate policies could positively or negatively impact factors such as energy use, transportation, food, and air quality. Students used online government databases to consider how the Paris Agreement affected atmospheric conditions over time, as well as the impact a country's food consumption (i.e., meat- vs. plant-based diet) might have on a variety of factors. They read graphic novels and biographies about individuals of color and those from low-income backgrounds in various climate change careers. The students acknowledged the experience helped them see themselves pursuing similar roles in the future.

Self-reflection activities were embedded throughout the course as a way for students to consider their identities as learners and guide their next steps in addressing a climate change issue directly related to their interests. For example, one student who enjoyed working on diesel engines investigated the type of adaptations needed to facilitate the efficient use of biofuels in these types of engines.

To learn more about sustainable growing practices, students took a field trip to a nearby farm that has engaged in these types of practices and had a virtual visit from a professor who had testified before Congress on this topic. Students also explored different cultures that depended upon insects as a valuable food source and went to the American Museum of Natural History to learn more about the critical role insects play in human health and agriculture.

A key focus of this unit was the development of leadership skills. Students were tasked with considering the intersection between their interests and climate change by exploring innovative ideas, managing obstacles, and developing actions they could take in their home, at school, in their community, or as part of their career pathway.

Students engaged in research about a climate topic of their choice and were tasked with finding someone in the field to participate in a career day at their school. After developing their own expertise on the topic, the students developed action steps related to it and shared them with other changemakers. For example, some students recorded TED Talks about their topics of interest and presented findings to their local mayor and borough council along with recommendations for change. Other students created websites and shared their findings with pertinent individuals for feedback. Some students reported that they appreciated the real-world applications of this unit, and many added their projects to their resume or used them to get internships.

Vignette 2: Sea-Level Rise

A middle school teacher who teaches in a community near the Atlantic Ocean used her students' concerns about recent news stories on the effects of sea-level rise on coastal areas to introduce a unit about the mechanisms behind and effects of sea-level rise, along with ways to take action. Leveraging the SubjectToClimate website, she downloaded lesson plans and a slideshow, classroom activities, and guidelines for leveraging online tools scientists use. All the resources on this website are free, vetted by scientists and educators, and include nationally recognized student learning standards. Working in lab groups, students considered what happened when they completed a series of actions designed to simulate sea-level rise. After recording their findings, the students engaged in a whole-class discussion about physical changes they see in everyday life and globally.

An interactive slideshow was provided to the lab groups to guide their investigation of melting glaciers. The activities in the slideshow were designed to foster students' development of many of the science practices incorporated into the NGSS. Students made choices about how they engaged in their learning experiences. They read articles, watched videos, and completed simulations at their own rate. They used Google Earth to "visit" several locations and observe how glacier melt and subsequent sea-level rise impacted the area. For inspiration, the students watched a video of activist Colette Pichon Battle, who discussed the impact of sea-level rise on extreme weather events. Students chose a community that would be impacted by sea-level rise to research, identified possible action steps, and then shared this information with other classes, local officials, or as part of a community night.

Vignette 3: Changing Bee Population

Third-grade teachers in Milwaukee[10] used a recent PBS news story discussing how the declining bee population relates to climate change to inspire an interdisciplinary unit that resulted in the school partnering with a local community group to plant pollinator gardens. In addition to addressing the NGSS, this unit was designed to help students develop their literacy skills.

The students learned about the importance of bees from several different types of resources. For example, the teacher utilized a beautifully illustrated picture book to help students who had initially expressed negative feelings toward bees have a more positive outlook on the contributions bees make to our ecosystems. A video was used to introduce the pollination process and explain the important role it plays in the habitat. Since these two resources depended on pictures to tell the story, students whose first language was not English were able to easily understand the concepts being discussed.

Students visited a pollinator garden at a local zoo where they met with a beekeeper who discussed how the bee population has been impacted by climate change. The topic of heat islands was introduced in child-friendly terms and students were able to ask questions about how Milwaukee was affected by the changing bee population.

Using an action-oriented approach, the teacher asked the students what they might do to help the local bee population. Students responded that they wanted to build a pollinator garden in their school. The teacher met with a local community advocacy organization to visit the school and develop a plan to aid students in planting a pollinator garden. The students also created posters describing their work to post in the school hallways. Students continued to tend to the garden and collected data on changes they observed.

Vignette 4: Water, A Precious Resource

During circle time in one early elementary classroom in San Diego, students expressed their concerns about the impacts the drought and more restrictive state water regulations[11] were having on their lives. To shift her students to a more solution-oriented mindset, the teacher decided the students might enjoy reading a book about how water can be harnessed for energy use. Because this topic was not part of her science curriculum, the teacher designed activities related to the book that would help her students develop key literacy skills described in the English language arts standards.

The teacher selected a book that highlighted the history and use of water wheels to read during literacy time. Realizing many of her students from low-income backgrounds might not have many books at home, she took great care to select a culturally responsive and inclusive book with beautiful pictures and that used language appropriate for young children. This was especially important for the students who were learning English and were not familiar with more academic language. The teacher modeled asking questions and defining new words with the aid of a glossary in the book that explained scientific concepts in terms appropriate for young children. After the teacher finished reading the book, the students were instructed to draw a picture of what they found interesting in the book. Working in pairs, students practiced their speaking and listening skills as they explained what was happening in their pictures and answered questions their partners asked.

To help the students better understand how the water wheels they read about worked, they worked in small groups to build their own water wheels using paper plates and cups. The teacher told the students they would take on the role of an engineer and would need to think critically to solve any problems they encountered as they experimented with putting different size rocks into the cups and spinning the water wheel at various speeds to witness how much power it could draw. The students displayed their water wheels at the school library, along with step-by-step instructions they had written about how the water wheels were constructed.

A highlight of the unit was a visit by a student's parent who worked at the local power plant. The parent answered the students' questions about what happened at the power plant and discussed some of the possible negative consequences of hydroelectric dams (i.e., ecosystem disruption). In addition, the parent talked about the types of careers available for individuals interested in how we can harness water as an energy source.

Questions to Guide the Selection of Resources

Climate change education requires the use of a wide range of resources. The questions below are designed to aid in the selection of resources that facilitate high-quality instruction about climate change. A single resource does not need to embody all the characteristics identified below. Further, educators can use these questions to consider how they might want to adapt their instructional methods to incorporate the characteristics listed.

CULTURALLY RESPONSIVE AND INCLUSIVE RESOURCES

Does this resource contribute to a learning environment that recognizes:

- Different identities and perspectives (e.g., racial, cultural, religious, gender, sexual orientation, age, political)?
- A student's preparation for learning may have been impacted by their socio-economic status or language ability?
- There are multiple entry points to learning based on a student's interest and ability level?
- There are many paths to understanding and multiple ways to demonstrate learning?
- The various cultural identities of the students and their communities in this space?
- The importance of resources in different formats, languages, and at different reading levels?

ACCURATE AND STANDARDS-BASED RESOURCES

- Does this resource provide accurate information?
  - For written and digital resources, is this resource credible? Who created this resource and what is its purpose? When was this resource created and who is the intended audience?
  - For individuals who serve as resources, how might this person's expertise or experience aid in students' understanding of the topic? Does this person have experience engaging with students of this age? Are there guidelines that address how students may interact with individuals outside the learning environment?
- In what ways might this resource be biased?
- Does the resource reflect the expectations of grade-appropriate student learning standards?
- Do the students have the foundational knowledge and skills necessary to use the resource (or can scaffolding be provided to aid in the students' understanding)?
- Does this resource contribute to an interdisciplinary approach to climate change education?
- Does this resource aid in students gaining essential skills necessary for college and careers?

RESOURCES FOR STUDENT-CENTERED LEARNING AND ACTION-ORIENTED ENVIRONMENTS

Does this resource:

- Provide an opportunity for students to engage in real-world processes or use authentic tools?
- Help students address an issue that is relevant and meaningful to their lives (e.g., an issue impacting their school, community, advocacy group)?
- Enable students to demonstrate their learning to an authentic audience?
- Promote an action-oriented approach to learning (offering solutions rather than focusing only on the negative impacts of climate change)?

Notes

1. Elaine Lin Wang, et al., "Teachers' Perceptions of What Makes Instructional Materials Engaging, Appropriately Challenging, and Usable: A Survey and Interview Study" (Santa Monica, CA: RAND Corporation, January 14, 2021), https://www.rand.org/pubs/research_reports/RRA134-2.html.

2. The New Teacher Project, *The Opportunity Myth*, 2018, https://tntp.org/publications/view/student-experiences/the-opportunity-myth.

3. Laura Kieran and Christine Anderson, "Connecting Universal Design for Learning with Culturally Responsive Teaching," *Education and Urban Society* 51, no. 9 (2019): 1202–16, doi:10.1177/0013124518785012.

4. Linda Darling-Hammond et al., "Implications for Educational Practice of the Science of Learning and Development," *Applied Developmental Science* 24, no. 2 (2020): 97–140, doi:10/1080/10888691.2018.1537791.

5. Wang et al., "Teachers' Perceptions."

6. Martha C. Monroe et al., "Identifying Effective Climate Change Education Strategies: A Systematic Review of the Research," *Environmental Education Research* 25, no. 6 (2019): 1–22, doi:10.1080/13504622.2017.1360842.

7. Darling-Hammond et al., "Implications for Educational Practice."

8. Monroe et al., "Identifying Effective Climate Change."

9. Sonya Remington Doucette et al., "Teaching STEM through Climate Justice and Civic Engagement," *Science Education and Civic Engagement* 15, no. 1 (Winter 2023): 6–16, https://eric.ed.gov/?id=EJ1388898.

10. New Jersey teachers may want to reach out to NJ Audubon (https://njaudubon.org/) or their Rutgers Cooperative Extension County office (https://njaes.rutgers.edu/county/) for resources about creating pollination gardens, challenges facing local bee populations, or addressing issues in urban heat islands.

11. New Jersey teachers interested in implementing this unit may want to research what types of governmental regulations impact water use in their communities.

CHAPTER 17

Bridging Classrooms and Communities for Climate Change Education

Tina Overman, Kelly Stone, Chris Turnbull, and Karen Woodruff

Collaboration between formal educational institutions and community organizations can be transformative for successful climate change education. By uniting the expertise and resources of formal educational institutions with the work of professionals in the broader school community, students may develop an applied understanding of the content and of how their everyday actions and decisions make a difference locally and globally. Communities are central in the lives of children, serving as the places where they explore the natural world and shape their identities through interactions with peers, family, and educators.[1] They are often centered on schools and geographically defined by school district boundaries.

Communities are also places where efforts to mitigate climate change are underway. Decisions made within communities contribute either to the exacerbation or alleviation of CO_2 levels. Understanding these localized initiatives enriches students' understanding of the broader climate-change landscape and empowers them to recognize the role they play within their communities.[2] Grounding learning experiences in the places where students live helps them understand that they can play a role in taking care of their place, whether they are in a rural, urban, or suburban environment.[3] Bridging the efforts of community members, informal education organizations, nonprofits, and others with formal education spaces provides opportunities for students to apply their knowledge authentically, grounded in the communities where they live, play, and learn.[4] As educators seek to enhance curriculum to include climate change education, partnerships with local community organizations are an asset. Through such partnerships, students are positioned as researchers,[5] changemakers,[6] and activists.[7] These interactions form the bedrock upon which children develop the attitudes and values that will accompany them into adulthood.[8]

As educators, we are each engaged in partnerships with local community organizations that enhance our teaching of climate change. In this chapter, we share vignettes of our collaborative work with community partners to enhance climate change education. We then make several recommendations for seeking partnerships and planning collaborative work in communities that will enhance instruction and provide a bridge for students to the everyday application of the important concepts.

Growing on your Property and Community Gardens

Tina Overman, Kelly Stone, Chris Turnbull

When teaching about climate change, the best place to start is outside! Look for available locations to lead a lesson, explore, create a new space, or test a design. Our first mission was to create an outdoor learning area in an empty courtyard full of mowed lawn space. If we want kids to feel connected to nature, the environment, and climate changes, it is up to us to get them outside.

Securing a greenhouse in our outdoor learning area created a direct climate connection to growing our own food, eating locally, watching plants' life cycles, composting, observing pollinators in action, and planting seeds and transplants. Again, the task might look overwhelming, so a great place to start is to reach out to a local farmer. Ask for their recommendations for the size of your space. What kind of greenhouse might work best? In our case, Jess Niederer, owner of the organic Chickadee Creek Farm, let us know she had an old hoop house she wasn't using and offered not only to donate it to us, but install it herself! We knew we needed one side to have doors for easy access, so again, we reached out to a local builder and parent from our school, Bear Tavern Elementary (BT), who was happy to help. The greenhouse extended our growing capacity and helped solidify our partnership with local experts who were happy to help.

After conversations with students and families about farming and environmental justice, we reached out to a local farm (Gravity Hill Farm), who supplied us with lessons, seeds, transplants, and ongoing education. That community link led to another when they connected us with both Rolling Harvest, a local food rescue, and our local Outdoor Equity Alliance. These new partnerships led to two different student presentations and a field trip to the farm.

When one class at BT was thinking about problems around our property, the students knew we used to have milkweed on our campus, and they were bothered by the fact that we didn't have much milkweed or many monarch caterpillars any longer. We thought about potential solutions, and they suggested

BRIDGING CLASSROOMS AND COMMUNITIES 225

Figure 17.1 Children tending a school garden. K. Stone.

we replace some of the mowed lawn outside their classroom with raised beds. As the discussions progressed, one student after another talked about how their parents could help.

Next step? Reach out to the parents! Soon enough, we had wood for raised beds, a group of parents to assemble and deliver them, and a truckload of soil.

Whether you have a small plot or a spacious meadow, growing plants for pollinators is an incredible way to support biodiversity and restore property. As we looked around the property for potential pollinator spaces, it became clear we had both big and small open spaces that could be used. We started with a small plot outside our gym area that had once been planted but was not very well kept. We used money from a fundraiser to buy twenty-five plants great for pollinators. Fourth-grade students planted them and then used them to think about the plants' structures and functions, as well as their connections to pollinators. The next step, creating a pollinator meadow, was a bit more involved. Again, we used funds from a fundraiser. We reached out to a local group to help prepare the soil for spreading the seeds. It has since grown into a space with abundant plants and grasses that have replaced more of the mowed lawn. This process has been a valuable experience for students to examine the grass and then step into a flourishing meadow to observe the difference in biodiversity.

Look for local grants to bring plants and flowers to your space. We received a grant from the Xerces Society, though we still had to work to get the space ready for over seven hundred plugs. We knew there was a farm run by a former student next door to our school. We were incredibly thankful to them for donating their time and one of their tractors to till the field. Every student in the school planted the plugs and watched the meadow evolve season after season. Countless lessons were led in the meadow.

Free trees are offered to third graders in New Jersey through the Forest Resource Education Center, so we give every third-grader a tree to take home and plant. The remaining trees are planted on the school property as another way to restore it. Planting trees year after year is yet another way that we demonstrate for the kids that this process is ongoing. Planting native trees led to lessons about the native wildlife that rely on the trees, carbon sequestration, and the benefits of trees. Getting teachers out in the field digging and planting helped us build community within the school.

Like with the outdoor learning area, we found that the best place to begin our journey toward a rich and effective climate change education program was to begin outdoors, connecting our students with nature. We knew our start would lead us to a beautiful unused and spacious courtyard as a newly protected space

Figure 17.2 Children observing a schoolyard pond. C. Turnbull.

where children could learn, explore, and engage in hands-on activities while growing and cultivating a bountiful garden, knowledge, and a love of nature's beauty.

In an area where a large percentage of the population's socioeconomic status makes access to health and nutritious fruits and vegetables difficult to provide routinely, we began with plans for a garden to support our students' nutrition. We formed a partnership with Providing Hope, a community outreach program that initiates youth and community programs focused on food, nutrition, and disease prevention. They work to improve New Jersey residents' health and quality of life, particularly for those facing significant life challenges. This partnership gave us a way to help fund the program and provide many students with life skills in planting and caring for a garden and access to a bountiful harvest.

In the planting, caring, and cultivation of a large variety of vegetables, our students were not only connected to a meaningful project, but they were learning about standards in each experience. From setting up garden arrays to measuring growth, predicting outcomes, observing, and journaling, students were engaged in the entire process. They learned that simply providing soil and water were not enough to sustain the thriving life a garden needed to produce a harvest. Weeds, sufficient space, soil conditions, water-versus-rain comparisons, helpful and pesky insects, and the brutality of a scorching sun in the middle of summer were all obstacles and lessons along this journey.

Plenty of obstacles and mistakes frequent the start of a new program or initiative and must be overcome. Some of ours continue ten years later, as we work together to find unique solutions to maintain the gardens. The addition of pollinator gardens, rain garden areas, rain barrels, and even weed barriers have contributed to the garden's productivity.

One of the biggest contributions to the outdoor learning space was the construction of a twelve-by-twelve-foot greenhouse that our facilities and grounds team built with members of the community and, of course, our students. Deciding the best greenhouse option was a difficult task, but the model that engaged students in the building process won, and they were part of a planning and construction process that opened many unequaled experiences. Students spent months collecting, washing, and preparing two-liter soda bottles to serve as walls for the enclosed growing space. Once approximately two thousand bottles were collected, construction of the space began. The bottles were held on bamboo sticks and connected to the greenhouse walls, providing a barrier from wind and insulating the space to maintain heat in cooler months.

In an area where hurricanes, nor'easters, and other storms are prevalent, we quickly found the bamboo sticks were not strong enough to withstand the wind. Students spent much time cleaning up and restacking the bottles. With other materials and the help of the facilities team, students found that PVC lengths were a better way to maintain the structure's strength. The number of problems that arise in this space facilitates so many unique opportunities for collaboration, conversation, debate, and problem solving, all of which are necessary for students to succeed in whatever paths they choose to follow as they grow.

As mentioned above, the greenhouse not only extended the growing season, but gave additional space to start seedlings and even store most of our garden tools. Our gardens have expanded slowly and have provided so many healthy options to families from our school and in the larger community. We currently grow a variety of herbs that both benefit the growth of other vegetables and keep pests at bay, in addition to vegetables selected and voted on by students and staff in the building. Through the growing experience, students are tasting new vegetables, like butternut squash, eggplant, okra, zucchini, squashes, and many more.

Farmers' Markets

Kelly Stone

With all the delicious and nutritious herbs, veggies, and fruits comes the concern of how to bring these harvests back to the community and, most importantly, share them with students. Transportation barriers brought challenges to making sure as many students and community members as possible could eat from the school gardens before the produce spoiled or overripened. With academic camps in progress and the help of local city officials, we secured a distribution spot at our local farmers' market. Many local municipalities host farmers' markets in their cities or towns, even sharing with nearby communities.

On a weekly basis, students in the summer enrichment and academic programs harvested, gathered, washed, bunched, weighed, and arranged the produce to bring to the farmers' market. Our district's transportation department provided bussing to transport students and the harvest to the park site where the farmers' market was held. While there, students arranged and presented what they had harvested to the community and charged minimal fees for their bounty. They counted and helped shoppers select vegetables and herbs while securing them in reusable bags to limit single-use plastic. Students were responsible for maintaining the stand, pricing the food, counting money, making change, and conversing with community members who came to shop. They discussed how they harvested and grew the products, as well as their favorites and how they ate them.

While they were waiting between customers, students were able to explore other shops around the market and select additional farm-grown items to bring back for tasting and comparison. Funds collected at the farmers' market helped make these purchases and gave us the opportunity to purchase additional seeds and seedlings for the next growing season. The real-world experiences students received were invaluable life lessons.

In addition to selling at the farmers' market, produce was harvested for students enrolled in the summer camps. Pop-up stations were presented at each camp location throughout the community. Students in attendance were able to shop and collect a variety of veggies and herbs to bring home to their families at no cost. This offered students the opportunity to understand how selections are made in a grocery store–type environment, with discussions on how these types of foods could be prepared at home. Students promised to sample the items they selected to broaden their palates and increase the number of healthy options for meals in their homes.

Additional produce was shared at both district and school events. We welcomed families into the school community at orientation days, where we provided them with samples of school-grown watermelon and cantaloupe and a bag of herbs and tomatoes to take home for dinner; this process also helped engage them in the culture of sustainability and climate-responsible students we foster in our building. Back-to-school nights offered parents and students the opportunity to shop together and plan and prepare meals grown at school, as well, and engaged our parents and families in learning activities throughout our building. Students shared with their families their excitement in harvesting during the day and preparing those yummy choices to find their way onto their dinner tables.

Composting

Kelly Stone

One of the most exciting lessons we have learned is that we can overcome all our obstacles right within our school community, sometimes with a little help from the partnerships we have made. Sustainable Jersey for Schools is one of the great partnerships that helps us teach our students about climate change and create opportunities for them to be environmental stewards. We applied for and were given a grant to begin a composting program in our school's cafeteria. The benefits are twofold: We are reducing waste from going to landfills while educating students, and we are using the compost or dehydrated food in our school gardens, enriching the soil and naturally supporting plant growth.

In the initial stages of the grant, we worked with Rutgers Cooperative Extension to audit the cafeteria to determine the amount of waste we produced daily. Although we had done this project with our students in the past, this audit was much more involved, detailed, and informative. We found we produced a total of 625.17 pounds of waste in three daily visits, or approximately 208.39 pounds of waste per day; 1,042 pounds of waste per week, and 37,510.20 pounds every school year—in other words, an alarming amount of wasted food. We then researched purchasing a composter that would best fit our building.

The ECOVIM 250 met the needs of the volume of waste we produced, and it is an indoor machine, making it accessible to students and staff regardless of the weather.

Located in the cafeteria, the ECOVIM involves all students in the entire process of converting food waste to compost for the school gardens. Food waste is collected and sorted daily to ensure there is no plastic or wrapper-type waste mixed in at a sorting station. The students are responsible for sorting their waste into appropriate bins. The waste is added to the composter at the end of each lunch period and if full, it is run overnight to produce compost. It is stored in large garbage cans through the colder months and added and tilled into garden beds in the spring.

Students and their families are invited into the courtyard garden in the spring to help prepare the beds, till the compost and soil, pull weeds, and prepare the beds for student planting. Partnerships among families help create a positive culture for the students and offer support for parents that may be experts in the field. The compost process has not only created a healthier garden and diverted 37,510 pounds of waste from going to the landfill yearly, but it has also encouraged responsibility and accountability within our students. They take ownership of the project and make changes and adjustments when warranted.

Living Wall

Kelly Stone

Imagine walking into a school foyer and having your senses come alive with the beautiful sights and sounds of nature. The beauty and functionality of a living wall connects not only our students and staff, but the community to the majestic beauty of the great outdoors. With much invested in creating a sustainability mindset and actions, our students and staff continually strive for new and innovative ways to grow in environmental practices and education. Our goal was to innovate, create, nurture, and maintain a sustainable space for all who enter our halls; thus, the living wall was built. To bring the vision to life and help fund the project, we collaborated with community partners from Sustainable Jersey for Schools and their grant program, as well as Project Sustainability in Minnesota—a little farther away, but no less connected to our community.

To ensure this living wall would have an immense impact on the school and local community, we consulted with a group of community members, including a doctor at our local children's hospital, a community health and social impact specialist, a climate change and environmental professor at a local university, and many teachers and administrators. Among the ideas discussed was how a wall like this one could connect the many cultures and communities that make up our beautiful city. From culturally diverse native plant species to community experts in horticulture, we believe this wall impacts thousands of children and their families. As a focal point for collaboration, the care of this wall provides not only meaningful and powerful learning, but a way to positively connect families, the community, and specialists at a time where social justice, equality, and equity are crucial.

The living wall in the lobby is used for many school and community events. In addition to afterschool programs, this area is an entryway for local sports programs, a local church, and other community programs that use the school after hours. The living wall is the focal point for welcoming members of the school community, local city community, and outside organizations, all while improving air quality and providing a social and emotional connection the beauty of nature for those who engage in the space.

Nature is an essential part of life, not only for aesthetic purposes or making us feel good, but also a way to connect and bring communities together to socialize, exercise, explore, grow, and learn. It is a significant connection to the way we live together. An article published in 2017 found that "Americans taking part in outdoor activities have been decreasing by 1% each year. Adults spend approximately 87% of their time indoors, a statistic which does not factor into the time they spend in their cars traveling."[9] Today's students spend an average of three to four hours on screens daily, based on statistics before the COVID-19 pandemic began. Factoring this into the time spent in school and in afterschool programs, there is little time to enjoy the beauty nature offers. Instead of plugging in to charge, we need to unplug and recharge ourselves.

The living wall brings the sights and sounds of nature into the walls of our school, inviting our students and families and the larger community to connect with nature, study and learn from it, and understand the importance of nature and the living world for all people, regardless of ethnicity or socioeconomic status. Nature should be a way for all people to come together, appreciate natural beauty, and learn.

Through the incorporation of the wall, our students can connect their learning and understanding with the real world while engaging in project-based learning across multiple subject areas. This project has applications beyond the classroom and connects our students to many potential environmental careers in their future. Understanding botany, horticulture, and agriculture encourages our students to consider many environmental jobs. Second- and third-grade students must meet life sciences standards that address plant and animal life cycles. This wall brings lifecycles to life and engages students in the purpose of identifying plants and including them as part of a living mural. They test plant species to see stronger versus weaker varieties that grow in artificial light or direct sunlight; they even test appropriate volumes of water against plant growth rates. First-graders can monitor, chart, and graph growth, connecting their learning to math and science. Bringing technical drawing into the lesson includes the fine arts for all grade-level students. The living wall connects to standards throughout school curricula across all grade bands.

Watershed

Tina Overman, Chris Turnbull

One of our many incredible community connections, and an invaluable resource, is our relationship with our local watershed organization, The Watershed Institute. This organization, like *your* local watershed organization, works tirelessly to protect and keep our water healthy and clean.

As storms become more powerful, erosion becomes a more and more significant problem. At our school, rushing water cut through our meadow after rain events, so we took this chance to create an opportunity. The students observed the destruction, asked questions about the problem, and then designed solutions. It was an easy connection for our fourth-grade students, because they were already thinking about some engineering design solutions to keep the soil and plants from washing away in the rivers cutting through. We also connected the issue to our local farmers, who told us they experienced the same difficulties with their crops. After we came up with some solutions and experienced successes and failures, we called in two community experts, Oliva Spildooren and Steve Tuorto from The Watershed. As they observed what was happening, they reminded us that solutions could be found in nature. We needed to plant the solution. Although the problem felt like a failure at first, there was great value in the lesson about nature-based solutions.

We were reminded that the best way to keep land naturally stable was to plant water-loving plants in that area. Olivia and Steve helped us determine some viable options to plan, such as a rush, like a juncus and a blue flag iris, both of which would help us naturally fight the erosion. Students in multiple classes created a plan for planting and set forth with the design. Later that year, we also added some pollinator powerhouse plants staff from The Watershed recommended.

At the same time, we were constantly battling unwanted plants, some non-native or invasive species, in our pollinator plot and meadows. The smaller plot gets heavily mulched each year. Again, while Olivia and Steve looked on, we learned yet another valuable lesson about the benefits of native ground cover. We knew just what to do next.

The Watershed helped us teach our fifth graders how to determine the health of our local waterways through StreamWatch Schools, one of their many education programs. Connecting with experts in the field, feet in the stream, engaged the students in a way that is hard to match in the classroom. Our students learned how to take a visual assessment of a stream, conduct chemical tests in the water, and collect macroinvertebrates. All data got synthesized to determine the health of the water and habitat in a specific location. It was incredible for the students to see all the issues that can negatively affect our water, watch first-hand how scientists test water, and then talk about how we might take action if needed.

Framework for Moving Forward

English philosopher and psychologist Herbert Spencer once said, "The great aim of education is not knowledge, but action." While the information in this book will undoubtedly challenge our thoughts around what is possible in the area of climate change education in our schools and will increase our understanding of the crucial elements of teaching climate change, the main objective is to both inform and inspire action in *your* community. Whether you live in Trenton or Tulsa or Tacoma, there are common threads that allow you to have an immediate impact and find success *where you are right now*. Every region is different, and every neighborhood, municipality, and county is unique, but with the framework below, you can leverage local resources to begin a movement that will inspire others, create a foundation for future growth and improvement, and most importantly, prepare students to be advocates and leaders in the area of climate change. Jeff Corwin, American biologist and wildlife conservationist, hit the nail on the head when he said, "I've always believed we will not protect what we do not love and cannot love what we do not know." By instilling students with information, passion, and tangible opportunities to create change, we ensure our students use their knowledge to create action that will help save our planet.

No work is too small or insignificant when it comes to climate change education, and simply creating a culture of empowerment, curiosity, and agency among our students is paramount. Once students know they can be true driving forces for change, anything becomes possible. Below are some suggestions for getting started:

- **Flexibility and an open mind:** Not one of our successes ended up the way we had initially planned it. If change is going to be lasting and dynamic, we must use the input, ideas, and resources others share with us (both children and adults). At Bear Tavern Elementary School in Titusville, New Jersey, what started as a school ground cleanup activity became a fifth-grader's idea for a nature trail, which led to a club, which turned into a full-fledged nature trail and outdoor classroom. The hundreds of tiny steps from the initial idea and completed project included adjustments, reflection, setbacks, partnerships, and feedback. When we truly let the students share ideas and lead the way, we generated excitement, community buy-in, and a true sense of purpose throughout our school. As adults and leaders, we had to figure out how and where to find the resources we needed to bring students' ideas to fruition. This included connecting with the local municipal works, borrowing tools from families, and even soliciting help from a local tree service. Education as a field often thrives on structure, but to truly let students lead the way, we must be able to relinquish some control, comfort, and our own vision of what "success" looks like. The success of our nature trail and the formation of the student self-named WACA (Wildlife and Conservation Association club) created excitement and buy-in and paved the way for countless future projects and successes.

- **Willingness to accept resources from a variety of sources:** Engaging activities and sustainable projects cost money and require materials. We have accepted donations, leftovers, excess materials, and expertise from a wide variety of sources in town and beyond, including local farmers, landscapers, our parent-teacher organization, our education foundation, former staff members, and so many others. Often, the available materials and donations, combined with flexibility and creativity, have led to incredible and sometimes unforeseen results, including a hoop greenhouse, meadow walkway made of leftover blue stone, and student-made benches made from composite decking pieces, not to mention raised beds and beautiful tile murals created from local Boy and

Table 17.1 Community and School Connections

Our connections	
County parks and land steward	Local Audubon Society
Local watershed	Parents (current and former)
Forest Service nursery	Former staff members
Township public works	Local food bank/food rescue
Local energy company	Local/state representatives for your town/district
Local farmers	City environmental commission and council
School education foundation	City farmers' market organization
Local builders and landscapers	Clean Ocean Action
Parent-teacher organization	Local restaurants
Local colleges/universities	Local hospital
Local authors	Project Sustainability: Living Wall

Girl Scout projects. See below for a comprehensive list of our local connections and partnerships.
- **Allowing students to guide and lead the program:** Students invest in a program where their ideas are heard and supported. They are the moving factor in gaining staff support and administration buy-in. Student curiosity and excitement builds culture throughout the school. It keeps them accountable and guides projects and innovation. For so long, intergenerational learning has been prevalent in both schools and familial structures. Students consistently learn from their elders through life lessons, examples, and prior knowledge. By creating an opportunity for students to become leaders, we reverse the roles. Youth become the experts and guide their peers, teachers, leaders, and families to make climate-conscious choices and engage in innovative projects to make real-world changes that benefit local and global communities.
- **Tenacity:** Rest assured; teams of people will be willing to give you countless reasons your ideas will not work. Be prepared to respectfully stick with your ideas, put in extra legwork, and navigate the red tape that exists in the education world. One of our biggest inspirations, Simon Sinek, said, "Innovators are more focused on the vision than the obstacles that stand in the way." There will be times you feel like you are swimming upstream because of a lack of support, scarcity of materials/funds, and general setbacks, but the most important thing you can do is look for a partner and keep going. While innovation and new ideas are always exciting, revisiting past projects also has tremendous value. Most student-led projects require yearly upkeep and problem solving, since they can become weathered over time. Tenacity is essential to keep the vision thriving, even if changes or adjustments are needed year after year. Having a team of committed teachers and students makes your project sustainable in every sense of the word.
- **Leaving room for joy:** This one is simple. The work you are setting out to do is *hard*, but it is also fun. Throughout your journey, it is so important to be present, mindful, and aware of the intangible impact you will make on the students in your care and your community. Whether you work with elementary students, are in higher education, or anywhere in between, the magic is all around you if you set the intention to find it. Let the journey bring you energy, excitement, and hope, because the work you are doing matters and makes a difference every day!

Notes

1. Ming Kuo, Michael Barnes, and Catherine Jordan, "Do Experiences with Nature Promote Learning? Converging Evidence of a Cause-and-Effect Relationship," *Frontiers in Psychology* 10 (2019): 305. doi:10.3389/fpsyg.2019.00305.

2. Martha C. Monroe et al., "Identifying Effective Climate Change Education Strategies: A Systematic Review of the Research," *Environmental Education Research* 25, no. 6 (2019): 791–812. doi:10.1080/13504622.2017.1360842.

3. Robert Barratt and Elisabeth Barratt Hacking, "Place-Based Education and Practice: Observations from the Field," *Children Youth and Environments* 21, no. 1 (2011): 1–13.

4. Kuo et al., "Do Experiences with Nature…?"

5. Amy Cutter-Mackenzie and David Rousell, "Education for What? Shaping the Field of Climate Change Education with Chidlren and Young People as Co-Researchers," *Children's Geographies* 17, no. 1 (2019): 90–104. doi:10.1080/14733285.2018.1467556.

6. Lynne M. Zummo and Sara J. Dozier, "Using Epistemic Instructional Activities to Support Secondary Science Teachers' Social Construction of Knowledge of Anthropogenic Climate Change During a Professional Learning Experience," *Journal of Geoscience Education* 70, no. 4 (2022): 530–45. doi:10.1080/10899995.2021.1986785.

7. Derek G. Shendell et al., "Knowledge, Attitudes, and Awareness of New Jersey Public High School Students about Concepts of Climate Change, including Environmental Justice," *International Journal of Environmental Research and Public Health* 20, no. 3 (2023): 1922. doi:10.3390/ijerph20031922.

8. Julie Ernst et al., "Contributions to Sustainability through Young Children's Nature Play: A Systematic Review," *Sustainability* 13, no. 13 (2021): 7443. doi:10.3390/su13137443.

9. Julie Beck, "Nature Therapy is a Privilege," *The Atlantic,* June 23, 2017, https://www.theatlantic.com/health/archive/2017/06/how-to-harness-natures-healing-power/531438/

CHAPTER 18

Nonformal Education about Climate Change

Pat Heaney, Allison Mulch, and Graceanne Taylor

The Nonformal Tradition

Nonformal, sometimes called informal, education is learning that takes place outside the traditional classroom. Nonformal settings run from aquaria to zoos and many places in between. A great deal of nonformal environmental education happens at nature centers and parks, led by naturalists and rangers. Many youth groups, often volunteer-led, also provide non-formal education. Scouts, Y-programs, and all types of camps are included in the category. Bell et al.[1] distinguish between three different types of informal/nonformal science learning: "informal environments" such as mass media, individual hobbies, or conversations; "designed environments," including museums, nature centers, etc.; and "programs" such as Scouting and other clubs. For our purposes, we will include the latter two, "designed environments" and "programs," under the general category of non-formal education.

The nonformal nature of these programs refers only to their settings being outside the schoolroom. The term should not be misconstrued to mean unprofessional or undisciplined, however. Indeed, many non-formal education groups promote rigor and excellence in their programming. For example, the North American Association for Environmental Education established guidelines for developing balanced, scientifically accurate, and comprehensive environmental education programs and materials.[2] Nonformal educators can draw from a variety of sources and be flexible in their presentation styles, while remaining scientifically accurate in their content. Because they are not constrained by classroom walls or the limits of a school subject, they can bring truly interdisciplinary, hands-on learning to their students. This experience makes many non-formal educators uniquely qualified to communicate about climate change. Drawing from the social sciences, non-formal educators often frame their discussions to suit their audience. Because they work with students of all ages, abilities, and socioeconomic statuses, they can use appropriate framing to lead activities and discussions. The National Network for Ocean and Climate Change Interpretation (NNOCCI) offers numerous training programs on using evidenced-based communication methods and providing the social and emotional support people need to engage as climate communicators.[3]

Non-formal educators may also be familiar with the "Six Americas" identified by the Yale Program for Climate Change Communication and George Mason University's Center for Climate Change Communication.[4] These six audiences were first identified using a large, nationally representative survey of American adults in 2008. The group has continued their research over the past fifteen years, outlining six different levels of concern regarding climate change. The most concerned are labeled "alarmed"; these are Americans who understand the human causes of climate change and are actively taking steps to mitigate it, and 2023 data show that 28 percent of Americans are in this group. The next group is "concerned"; the 29 percent of Americans in this category believe climate change is a threat and support climate policies, but they see it as a less immediate, far-off threat. Fifteen percent of respondents are "cautious," meaning they haven't made up their minds about the reality of climate change. The next group, "disengaged," has very little awareness of the issue; about 6 percent of the population is identified this way. Eleven percent of the population is "doubtful," meaning they don't believe climate change is a real issue, and another eleven percent are "dismissive"; this group not only denies climate change, but might also believe it is a conspiracy or hoax. The researchers in this study also created a short version of the survey and a group tool so educators and others can see where their members fall on the scale. Using the Six Americas framework, presenters can then frame their message to effectively communicate to their audiences.[5]

Since many non-formal educators are focused solely on communicating climate change, they have the capacity to take part in targeted courses and training. Geiger and colleagues observed over one thousand presentations by non-formal educators at over one hundred different science learning centers, both before and after undergoing training on framing climate change.[6] They found audiences had a greater understanding of climate change, more hope, and increased intention to engage in community actions after the presenters were properly trained in framing climate change.

Climate Programs at The Watershed Institute

The Watershed Institute is a nonprofit agency in Pennington, New Jersey that has worked to keep water clean, safe, and healthy since 1949. They employ experts in conservation, advocacy, science, and education, and their educators have developed climate education programs using resources from diverse sources. The Climate, Health, and Justice lesson plan started with a modeling exercise the American Public Health Association (APHA) provided in which high-school students read information from the National Institute of Environmental Health Science, then work to fill in a model displaying the climate driver, environmental condition, hazard, health condition, and vulnerable populations. The APHA lesson was developed for classroom use employing PowerPoint slides. Health professionals were asked to take the lesson to high schools. After attending APHA training, Watershed Institute staff worked to make the lesson more kinesthetic and engaging. With APHA's permission, they took the lesson off-screen and wrote the components onto wooden craft sticks in different colors. Students work in teams to find the correct components to construct their models. The modeling activity is followed by a portion of the Vector-Borne, Waterborne and Heat-Related Illnesses activity from Project WET's *Water, Climate, and Resilience Guide*.[7] In this activity, students "diagnose" one another with climate-related health conditions. The program ends on a hopeful note, with students profiling young people around the world who are doing positive work regarding climate, health, and justice. This program can take place outdoors or indoors and is adaptable to a variety of time constraints, group sizes, and other conditions.

For intermediate and middle school students, Watershed Institute staff adapted a bird migration game and created an exciting active model outlining components of climate mitigation and adaptation. The Climate Challenge, an active, experiential learning game, teaches students to navigate our changing world. The activity introduces the concepts of climate mitigation and adaptation by providing a glimpse of possible futures. Learners roll oversize dice and move to stations that require them to address various scenarios, such as mitigating climate change, adapting to it, or confronting unexpected problems. Throughout the game, students keep track of what happens to them and evaluate the consequences of their decisions. The activity wraps up with a lively discussion of what happened during the game and how the results can inform their choices to be part of a solution to climate change. Teachers can extend the discussion in the classroom by assigning research about the scenarios they encountered along the way. The game can be set up indoors or outdoors, with the field of play ranging from a classroom to a giant field or woodland trail.

Climate Change Education in Scouting

Girl Scouts of the USA (GSUSA) is the country's largest non-formal education organization dedicated solely to girls. Part of their mission is "to build girls of courage, confidence and character who make the world a better place.[8]" Indeed, the Girl Scout Law by which members promise to live includes using resources wisely, being responsible for what one says and does, and making the world a better place. GSUSA currently has two Climate Challenge programs, one for grades K to 5 and the other for grades 6 to 12, offered free of charge on their website. Each program is divided into three sections: "Explore climate science," "Connect with your community," and "Share hope." Girls are encouraged to complete ten activities, including creating a challenge project and sharing it with their community, to earn a patch. These are offered by the national organization, but within the organization are many regional councils, some which offer their own climate change badges and programs.

The Boy Scouts of America (now called Scouting America) is a nonformal education program for both boys and girls. According to their website, they provide the nation's foremost youth program of character development and values-based leadership training. Although they do not have any specific badges or programs on climate change, it is briefly mentioned in the Conservation Good Turn certificate program.

Educators in nonformal settings, especially in public-facing venues like museums and zoos, can reach a great number of people. They are also able to collaborate with formal schoolteachers to enhance classroom learning. Our next section focuses on these collaborations.

Collaborations to Enhance Classroom Learning at New Jersey Audubon

Local and national nonprofit organizations host climate change programs that rely on highly skilled education staff and strong collaborations. New Jersey Audubon (NJA) is a 126-year-old nonprofit organization connecting all people with nature and stewarding nature today for all people tomorrow. Climate change is a direct threat to nature and people, so their education team collaborates locally and nationally to solve this problem through education. The following collaborations were created through this work.

As an affiliate of the National Wildlife Federation, NJA hosts their national programs locally with full-time education staff dedicated to providing schools with programs, resources, and professional learning that support climate change education. EcoSchools U.S. is the longest-running program providing schools with student-driven, solutions-based learning opportunities that support sustainability and climate action education. It was recently redesigned to encourage more community collaboration through bite-sized projects. Students are empowered to act on the climate challenges facing their communities. NJA collaborated with Arts Ed NJ to customize EcoSchools U.S. actions by creating opportunities to investigate and represent climate data through the arts so participants can better understand, visualize, and reflect on the actions and factors that impact habitats, wildlife, and people. Student learning is further deepened by practices from New Jersey's Arts Education and Social and Emotional Learning Framework to incite hope and solutions through student artwork.

Furthering their collaboration with the National Wildlife Federation, NJA now hosts the Resilient Schools and Communities (RiSC) program in New Jersey. In the pilot year, teachers from Wildwood High School participated in summer professional development with teachers from Coney Island, New York to learn about climate change impacts and large-scale mitigation efforts led by the US Army Corps of Engineers. They also heard from community advocates fighting for climate resiliency, met local artists who stirred emotions and depicted the urgent need for action, and worked in teams with fellow educators. Now in its third year in New Jersey, RiSC supports a cohort of teachers who meet biweekly, implementing and evaluating place-based curricula. Through field trips, students explore impacted coastal areas to make observations, collect data, and create solutions to support their communities.

Through an activity designed to develop community connections, students from Wildwood High School interviewed city officials, community leaders, and business owners to create video diaries documenting their memories of loss and destruction from Superstorm Sandy. Students then shared their own memories of a more recent storm. They explored local vulnerability reports and created a bilingual community vulnerability survey that they distributed and analyzed to determine the community's needs. They learned that 69 percent of respondents were concerned about the impacts of climate change, but only seventeen percent were aware of the flood-mitigation efforts within their communities, and only six percent of respondents had ever participated in a community awareness event. Based on this data, students created bilingual emergency preparedness brochures and presented the information and brochure to their school board.

For a related field trip, the Atlantic County Municipal Utilities Authority (ACUA) provided tours of the Haneman Environmental Park's landfill and recycling center, where they discussed impacts from major storms on the landfill, such as the overwhelming construction trash created when homes were impacted by storms. The trip also included a tour of ACUA's wastewater treatment plant. Located along the shoreline, where saltwater intrusion affects the structure, the bulkhead is too short to provide enough protection during large storm events and results in wastewater overflow. With a focus on solutions, students also visited the facility's wind farm and solar array. Then, students from Wildwood took a walking field trip to their beach; used current predictions of sea-level rise to mark how far inland water could reach; and assessed how that flooding might impact habitat, vegetation, and manmade structures.

All three RiSC schools visited NJA's Nature Center of Cape May for a culminating event at the end of the school year. Students put on waders and pulled a seine net through the waters of Cape May Harbor to see firsthand the species living just feet away from the beach. Environmental educators engaged the students in discussions about population dynamics and how different species may be affected by sea-level rise and warming water temperatures. Students also examined live horseshoe crabs in the aquatic lab and learned about the ecological benefits of horseshoe crab eggs and the value of horseshoe crab blood to the medical world. They also discussed how these "living fossils" could adapt to a changing climate. Student teams formed model dunes and tested their strength against a surge of water to witness firsthand the risk owners of waterfront homes face, especially with more frequent and severe tropical storms.

Wildwood High School displayed students' public educational art during the event. This moveable art was also displayed in public areas, including at the school, in the municipal office, and on the boardwalk. Egg Harbor High School and Cape May Vocational School students created and shared informational brochures and educational climate change games they had created as a form of community engagement. Guest speakers included North Wildwood Mayor Patrick Rosenello and Wildwood Mayor Pete Byron, who gave insight into the beach erosion in North Wildwood and the local and state responses reported in the news. During the 2023–24 school year, the addition of a sculpture artist worked with students to create and install climate art in their communities.

In a collaboration with Climate Generation, NJA provided a three-day summer institute on climate change education for teachers. It included two virtual workshop days with educators nationwide followed by an in-person cohort day for New Jersey teachers featuring climate education presentations from multiple organizations. Sustainable Jersey for Schools gave an overview of their NJ Student Climate Challenge and resources provided by SubjectToClimate. OASIS (Organizing Action on Sustainability in Schools) introduced its networks of public and independent schoolteachers, and a trainer with the NNOCCI introduced the framework for communicating climate science. NJA's Garden for Wildlife intern spoke about impacts to New Jersey's habitats and provided an overview of phone apps that can be used to host a school or community-wide bioblitz. For hands-on activities, teachers explored climate impacts by investigating water absorption, heat mapping, and carbon sequestration. Rutgers University also provided overviews of climate data and engaged teachers in designing and building wind turbines. The College of New Jersey led teachers in creating models for exploring the reduction of wave forces by constructing and testing artificial mussel beds.

Schools and informal education providers have complementary strengths when teaching about climate change.[9] The collaborations described here, EcoSchools U.S. and RiSC, are examples of how we can leverage these strengths to create climate education that is solutions-focused and provides opportunities for student agency. By working on facets of climate change that impact their own communities, these partnerships help prepare students to act in the future. As detailed in Chapters 1 and 3 of this book, many state and national organizations, as well as teachers from local schools, collaborated to create the New Jersey SubjectToClimate website, merging formal and nonformal education settings for the greater good.

Helping Everyone Contribute to Climate Change Solutions through Informal Learning

Informal environmental educators speak with "everyone" every day. Since they often work in places where the general public visit, they are challenged with communicating to a broad audience. In this section, we will discuss this challenge and suggest how informal educators lead everyone to climate change solutions.

Environmental educators must be both communicators and scientists. It is critical that they connect with people. Their task is to empower community members to realize their own capabilities on the path to solutions. If they accept that most people are doing their best with the resources they have, they can take an empathetic approach to this work. Realizing everyone's strengths, interests, resources, and skill sets is the key to successfully adapting to climate change. Solutions will come from all sectors of our community.

Earth's systems, bacteria, animals, fungi, and plants are all infinitely evolving around us. Educators know ecosystems exist in a delicate balance. All living things have a special interest in making sure the systems within which we live stay habitable for our species. Humans, however, have a special ability to drastically alter that balance. The pace of the changing climate is a result of human influences on these systems, which means people have the power to change their behaviors and slow the negative impacts of a changing climate. Informal educators must help others understand that protecting and conserving earth's systems will result in clean air, clean water, and healthy people. People should be shown examples to build the confidence that they have power to make change.

The topic of climate change can elicit strong emotions in certain people. In some situations, it can be added to the list of taboo dinner-table topics, like religion and politics. However, to make change, we must address the issue. The outcome of anthropogenic climate change is playing out before our eyes and as a result, people react in different ways to cope and survive. Informal educators have a unique ability and burden to reach all the audiences other professionals are not reaching.

Scientists speak to other scientists and sometimes to activists and advocates. Advocates speak to elected officials, policymakers, and other decision-makers. Formal educators speak to schoolchildren and adults who self-select to be in school. Informal educators (especially in close-knit communities) can be the link between these professionals and the public.

The public includes many groups: people who are young, elderly, and in between; those who are Democrats, Republicans, and Independents; individuals who are knowledgeable about climate change, as well as folks that are hearing about it for the first time. The public includes people of different races, ethnic backgrounds, socioeconomic statuses, and a wide range of life experiences that influence a person's perspective. Somehow, an informal educator must take all the information from scientists, advocates, stewards, and formal educators; digest it; and help the massive public understand it. They must meet people where they are so they might change their behavior in favor of a more sustainable future. So, how do they communicate about climate change to literally anyone they encounter at programs and facilities? Let's go back to empathy. Their first task is to listen and observe so they might understand a person's experience. What is the person saying? What are they wearing that can help indicate their interests? Who are they with? Are they ready to listen? Of course, an individual is much more than their clothes and outward appearance. However, in an informal setting, educators often don't have time to get to know someone completely before they need to communicate with them. Certainly, rambling facts and information is one way to attempt communication, but it is not likely to produce the behavior change we are aiming for.

Once a person is engaged, informal educators can practice active listening. Does this person depend on the land as a hunter or fisher to directly feed their family? Is this a commercial fisherman or farmer that must feed the community? Is this person a policymaker tasked with serving the community? Does this person have a family or cultural tie to the land and water? Both Indigenous people and vacationers can have a familiar or cultural tie to a place. Is this person expressing a need that a more sustainable solution might help satisfy? All the above individuals have common characteristics. They all care about the land and water, but they have different reasons to connect to the land, the water, and their community. This can make the job of an informal educator even more complex.

Educators must understand the nuance of climate change impacts and phenomena to educate the community, no matter what angle they might need to approach it from. For example, consider a homeowner who vacations at the New Jersey shore. An informal educator might ask them to describe flooding they have experienced near their shore home. By asking a person to share their experience, educators cut out the need to make them "believe," thereby validating their experience and creating trust. This approach requires listening and showing care about the homeowner's experiences and needs. One sure way to unravel all that trust is to unload all the facts about *why* that flooding happens. The facts and figures of climate change are complicated and can be overwhelming. Feeling overwhelmed and anxious can lead to apathy, and apathy doesn't lead to change. Similarly, distrust results in a lack of positive behavior change.

For some educators, the more they understand climate change, the more emotional they feel and the more they want to do something about it. Naturally, they worry about our future. Instead of burdening someone with urgency, we can do better by leading with solutions and creating a trusted bond with a person to continue that relationship. It is important to steer clear of perfectionism, because it is impossible for every person to do everything needed to reduce climate change's impacts. Informal educators must focus on what people can reasonably do and lead the conversation with hope. Gateway experiences are excellent ways to engage people and meet them where they are. These experiences are easy to attend and can create community and instill hope with a collective effort. Trash cleanups, microplastic sifting events, or rain-barrel workshops are great ways to get people involved, create relationships, and build a community to combat the larger challenges of climate change.

We sometimes label people as those who care and those who don't care. At the core, whether they know it or not, every human needs clean water and clean air to survive, which means they have a need to address. Educators must practice empathy for all members of their community, even for people with whom they do not initially agree. How can they listen to a community's needs as well as an individual's needs? How can they build trust so that a five-minute conversation leads to a decades-long working relationship? Once they earn that trust, how can they empower individuals and communities with hope and portray real actionable solutions that translate to behavior change?

One way to communicate well with people of different backgrounds and interests is to choose words purposefully. Teaching about local ecosystems is a great way to help local people connect with these topics. MyCoast offers a platform for individuals to share photos and text about New Jersey's changing shorelines and can be used to help spark discussions about the reasons for this change.[10] But the words we choose to talk about changing shorelines also matter. "Climate change" and "global warming" have different connotations for different audiences. The same is true for "sea-level rise" and "coastal flooding." Political strategist Frank Luntz wrote a text, *Words That Work: It's Not What You Say, It's What People Hear*,[11] suggesting that individuals consider word choice very carefully. He advocated for using the term *climate change* rather than *global warming* more frequently in news media, and this change shifted the conversation across the US, perhaps even softening the issue for some.

Marketing and communication experts focus closely on word choice throughout their work. Often environmental educators come from education or science backgrounds, and they rarely have degrees in communication. They tend to learn about communication through experience or professional development. It is a skill they need to practice if they want to explore present-day climate challenges. Anthropogenic climate change can certainly lead to the human race suffering and struggling. Patience and the ability to understand and communicate with one another will determine success in overcoming these challenges. In the informal sector, educators have significant capacity to make that possible as they cultivate spaces where they can practice empathy, listening, trust, and instill hope to address climate change. To build some tools and strategies for facilitating productive discussions about climate across sectors, consider the Connecting on Climate tools published by EcoAmerica.[12]

With these tools, the examples presented above, and of course, empathy, non-formal educators can continue to make change and work toward a resilient future.

Notes

1. Philip Bell et al., eds. *Learning Science in Informal Environments: People, Places, and Pursuits* (Washington, DC: The National Academies Press, 2009).

2. Bora Simmons, "Guidelines for Excellence Series," North American Association for Environmental Education, September 29, 2022, https://eepro.naaee.org/resource/guidelines-excellence-series.

3. National Network for Ocean and Climate Change, "Training Opportunities," https://nnocci.org/Training/. Interpretation.

4. Anthony Leiserowitz et al., "Global Warming's Six Americas, Fall 2023," Yale Program on Climate Change Communication, December 14, 2023, https://climatecommunication.yale.edu/publications/global-warmings-six-americas-fall-2023/.

5. Breanne Chryst et al., "Six Americas Super Short Survey (SASSY!)," Yale Program on Climate Change Communication, 2024, https://climatecommunication.yale.edu/visualizations-data/sassy/.

6. Nathaniel Geiger et al., "Catalyzing Public Engagement with Climate Change Through Informal Science Learning Centers," *Science Communication* 39, no. 2 (2017): 221–49. doi:10.1177/1075547017697980.

7. Project WET, "Climate," https://www.projectwet.org/programs/climate

8. Girl Scouts of the USA, "About Us," accessed May 28, 2024, https://www.girlscouts.org/en/discover/about-us.html.

9. Roberta Howard Hunter, Gail Richmond, and Eleanor Kenimer, "Informal Science Educator Professional Identity: Perceptions of NGSS, Work with Teachers, and the Centrality of Place," *International Journal of Informal Science and Environmental Learning* 3, no. 2 (2023): 1–28. https://www.diopress.com/_files/ugd/b89959_b9878501792e4f5f862029eda13d2f00.pdf.

10. https://mycoast.org/nj.

11. Frank Luntz, *Words That Work: It's Not What You Say, It's What People Hear* (London, UK: Hachette UK, 2007).

12. Ezra Markowitz, Caroline Hedge, and Gabriel Harp, *Connecting on Climate: A Guide to Effective Climate Change Communication* (New York, NY & Washington, DC: Center for Research on Environmental Decisions at Columbia University & ecoAmerica, 2014), https://ecoamerica.org/wp-content/uploads/2014/12/ecoAmerica-CRED-2014-Connecting-on-Climate.pdf.

CHAPTER 19

"Mom, What's for Dinner?" Climate Action Served with a Side of Hope

Helen and Grace Corveleyn

Our Climate Journey: A Mother-Daughter Perspective

As the climate crisis becomes more of a center point in our world, it is no longer a subject we can ignore in the media, in the classroom, or at home. We are faced with a multitude of symptoms of a warming earth and must be armed with solutions and ideas to tackle this crisis head-on. We are surrounded by this messaging in our daily lives and will continue to hear it for the foreseeable future. As a mom and an educator, I have tracked the climate situation in both formal and nonformal education settings, through field research, in large-scale conservation projects, and at home as I raise my family. When teaching young children, you get used to simplifying matters to balance truth and accuracy. Using the best scientific facts available, explaining the climate crisis to little learners boils down to this: Any pro-environmental behaviors you can incorporate into your life will help the climate. It is all related and interconnected. On my journey of educating others and being a planetary steward myself, I realized being an environmentalist *is* being a climate warrior. Some days my battle cries are louder than others. Some days, I make faster, more efficient progress toward a better outcome for our earth. Other days, I need to restore my spirit. On days where the work seems too big and the difference I'm making is too small, I turn to my daughter Grace for hope, community, and connection.

We thought we would share with you the journey we continue to walk together as an example of how to demonstrate climate support and intergenerational connection. In this chapter, we explore methods of promoting environmental behaviors in a family setting, climate-positive changes you can make in your household to promote awareness and change, the experience of ecoanxiety from different generational perspectives, and how to become active in climate advocacy. Grace and I asked each other questions that may offer insight into dialogue between generations.

Developing Pro-Environmental Behaviors

Helen: Grace, how did you develop a love for nature? Where did it come from? Did the time spent in nature drive you toward climate action?
I think my passion for climate action also developed specifically from my love of the ocean. From a very young age, I was obsessed with the feel of the breeze, the captivating sparkle of the water, and the way the ocean always brought me a sense of peace. Growing up and spending time on the beaches of New Jersey and then seeing my favorite beach trashed disturbed that peace. It ripped me out of my little happy bubble, and it made me upset. It made me realize beaches needed saving. So, I started to do research about human-created waste and how it affected ocean biomes and their surrounding areas. The research was alarming; it made the twelve-year-old Grace sad people didn't see how wonderful her favorite place was. When they looked out at the ocean, they didn't see peace and beauty like I did. What they saw was a place for trash and their own personal edification. At that point, I started thinking, "If this is happening at my beach, what is happening at other beaches? Are they experiencing the same conditions?" That question led me to the discovery that this problem was so much bigger than I had thought. On a fourth-grade field trip to the New Jersey beach of Sandy Hook, I discovered the amount of trash on my favorite beach further up the coastline paled in comparison to how much trash I saw that day. My classmates and I spent two hours walking on the sand and picking up trash; we finished the day with fourteen large trash bags full of the waste we had collected. Looking out on everything we had picked up, I turned to one of the Sandy Hook volunteers and asked, "Why are people doing this? How can we stop it?" She laughed and replied that I should look into a career in ocean conservation. And that's where everything fell into place. I realized this could be my job. Saving the ocean, being on the beach, and teaching people how to clean up after themselves could be *my job*. That is eventually what triggered my need to be in the climate space. Knowing there was both a problem in the natural locations that had always brought me comfort and an opportunity to fix them gave me the drive to get started and the hope that a better climate future could develop from my actions.

Grace: Mom, how did you communicate your passion for the environment to us from a young age? What factors influenced your eagerness to teach us about the environment? How did you model environmental stewardship

for us in the hope we would follow your lead? How did you change your approach on environmentalism when talking to your students versus talking to your kids?

Growing up, I lived in a family that prioritized spending time together in nature. My parents loved to travel and share new places with us. Always a mix of culture and nature, they brought us to incredible state and national parks, with my dad always prioritizing finding an "old-fashioned swimming hole," which meant this perfect out-of-the-way place where we parked on the side of the road, giggled in our bathing suits, and nimbly danced over rocky paths to find cool clear waters to play in. My mom took on the role of finding museums and off-the-beaten-path oddities perfectly curated for our age groups, linking us to the culture of the area. It was these early experiences that created the basis for my desire to find pockets of our earth to love. I know now that "place-based education" is a method of connecting learning to the places you experience both in and out of the classroom. These early experiences of traveling in national parks rather than going to resorts or theme parks (although we did do bits of that, as well) drove me to always have a sense of "wanderlust" that I felt I wanted to impart to my own kids.

The desire to constantly explore has inspired many intergenerational family vacations, and when I consider the threat the climate crisis has on those preserved lands, it becomes very personal to me. I've been a part of the environmental movement since the age of thirteen. I wore pins on my jean jacket with slogans like, "Stop dumping toxic waste," and "The earth belongs to our children." It was from those early experiences in nature that I knew my life calling would be to do anything I could to preserve these spaces I loved exploring. As an environmental policy major in college, I took every opportunity I could to research the Kyoto protocol, which was the earliest call in my climate journey for greenhouse gas emission reduction in 1997, so much so that one of my professors pulled me aside and said, "Helen, you can research other areas. This is the fourth assignment that you've written about Kyoto." But even then, I was drawn to how to divest from fossil fuels and how to convince others to follow suit. We were in the middle of the ozone crisis at that time, as well, and I was on a personal crusade to stop everyone I knew from using aerosols of any kind. I was able to convince family and friends to be more aware and reduce their consumption, and that planted the seed of hope in me. Witnessing behavioral change is something I began to value. With every intention of working for the federal government in environmental policy, I began student-teaching to complete my second major of education and taught high school environmental science. It was with these students I recognized that growing a generation of environmentally literate students was more my calling than working with adults in state or national government. Shocked that this would be my path, I became a middle school science teacher and *always* taught through an environmental lens.

From the early 2000s, the climate movement ebbed and flowed while the ozone layer healed. I worked with students to educate around the major environmental trends of the time: recycling, keeping the oceans clean, preserving endangered species, and protecting biodiversity. With each of these topics, greenhouse gas emissions would roll through, more in looking at renewable resources than at global warming, but as my family grew and the world progressed, I watched as storms became more erratic and my family lived through devastating New Jersey hurricanes. The heat in the summer became more and more unbearable and my beloved snow days (forever a school-aged girl crossing her fingers in the winter for a day off school…) became more of a distant memory than an event that happened multiple times per season.

As these experiences became more regular, so did climate conversation in our family. Environmental action became a pillar in our family values. When we worked on Scout projects, they were conservation-based. When we were together in national parks, we marveled at preserved lands and talked about why it is so important to prioritize that in our political system. Our family integrated gratitude for clean air and clean water. We made sure we didn't take our experiences in nature for granted, and in turn give back through volunteer organizations, community cleanups, and climate activism.

When I think about the difference between raising a family with environmental values and teaching environmental values, the two go hand-in-hand. But my duty as a teacher is to bring students to this conclusion on their own by presenting facts and letting them make their own choices about how they will act. As a mom, I expect my children to respect and value the earth because it is part of our family's mantra. I sometimes must check myself when I am in the classroom to make sure students know they have their own decision-making power when presented with scientific facts. At home, we are by no means a perfectly environmental family. There is always more we can do. While as a family, we prioritize taking steps to reduce our carbon footprint through investing in solar panel technology on our home, composting, growing food, and purchasing an electric car, there is always more work to be done. And then the lines of teacher/mom become blurred. Always wanting to learn more and educate others, my family will always be my first students.

Fostering a Climate-Positive Family Attitude

Helen: Grace, did you ever feel pressured to take up a cause because I felt passionate about it?
I get this question a lot, and the answer is always yes and no. I do feel I got my start in the climate space because of you, but the work I choose to pursue and the

issues I care about are always driven by me. I see how happy and empowered the climate space has made you, and I chase that similar feeling but don't ever feel I need to start something because you started it. When I mention my work in the climate space, people always ask me, "Did your mom set you up with this?," and the answer is always no. I seek and apply for opportunities on my own and do my own work in the climate space to build my own reputation as an environmentalist, not because I have ever felt pressure from you to do so. However, I do feel my interest in the environment makes our relationship stronger, and that is something I am very grateful for. It gives us a connection that you don't have with anyone else in the family, so I enjoy having that piece of you all to myself. Had I not had you to inspire me to get involved so early in life, I would not be as passionate about climate as I am today. I look forward to continuing your legacy in the climate space and hope that someday, you'll be proud of the way I chose to forge my own path in the environmental movement and use you as my inspiration to make change, just like you always have and always will.

Grace: Mom, do you feel like I'm copying you by pursuing similar interests?
Hahaha! Never! After teaching for nearly twenty-five years, I've always felt (selfishly) I want the very best and brightest of my students to fight for our planet. The combination of being a good, clear communicator and having that backed by a respect and a talent in STEM is a powerful combination. I see that in a handful of students each year, and it is exemplified in you, Grace. I am a very proud mom that you are using your talents for our climate and our earth. When we are together enjoying a sunrise over the ocean in Baja or in Acadia National Park, I feel a spiritual connection to our surroundings that is amplified by sharing it with my daughter. I also have a profound gratitude that you look through the lens of stewardship, and when you appreciate natural beauty, it elicits the desire to fiercely protect it in the same way. That provides me with a deep satisfaction and a comfort for our future. As a mom, I want you first and foremost to be happy in your chosen career. I know that if being surrounded by the natural world is incorporated into your job, happiness is a likely outcome. Additionally, I know you are filled with the ability to create change. You are a natural leader and an incredibly insightful thinker. STEM is your passion, but your ability to write and communicate will make you successful and necessary in the climate conservation space.

Eco-Anxiety

Grace: Mom, how do you feel adults are handling eco-anxiety? As an adult, do you feel there is a space for you to talk about these feelings?

I don't think adults talk about this at all. I read articles about it and will talk about it in academic settings, but often, I feel like a lone wolf in my eco-pursuits as an adult. I absolutely feel eco-anxiety. I also feel alone in this arena in my everyday life. I have a few friends who I can talk about it with and mention it before I see the inevitable "zone out" because they are not working within the confines of conservation. Luckily, I do have a platform in the classroom and with students and colleagues in the Green Team after school, which allows me to be able to talk honestly to students about our feelings of overwhelm, the uncertainty of our future, and how we can bring hope to others. The act of talking, teaching, and sharing eases some of the anxiety I carry with me.

One of the great joys in my life was pursuing a graduate degree in conservation education in a wonderful program called Project Dragonfly at Miami University of Ohio. This gave my environmentalism a clean breath of fresh life and provided a community of like-minded people for me to spend time and learn with. After receiving my degree, I had the great fortune of returning to teach with Project Dragonfly, a program I fiercely believe in. During unique international field experiences, groups of conservationists gather to study and feel the camaraderie and support of people who are fighting for our climate, our land, and our water. This became my climate space, and while we don't explicitly talk about climate anxiety all the time, there is an understanding and connection to share frustrations and ideas, which in the end reduce the anxiety connected with the health of our planet overall.

Helen: Grace, how do you balance your feelings of eco-anxiety with action? Do you feel like eco-anxiety is a real issue among the middle school and high school population?

As a young person in the climate space, eco-anxiety is something I frequently think about. Questions like, "How are we going to combat fossil fuels? How will we stop drilling for oil? What about plastic pollution and food waste?" are ones I think about almost every day. As I think about these questions and they begin to cause anxiety within me, I have to remember I can't let myself dwell in the feelings too long. That's because stressing about these issues and sending myself into a spiral won't accomplish anything except make me more tense. Instead, you have to foster the emotion you feel in your anxiety into doing something about it. That's where most of my activism stems from. You can't just sit around and be stressed; you have to use how you feel and *do* something about it. Once I catch myself in this behavior, I start thinking about what I can accomplish in that precise moment to make myself feel I'm using that emotion to empower myself and others. Reposting climate change articles and pages to my Instagram story, brainstorming ideas to bring to the next meeting of my school's Youth Environmental Society or promising myself to not buy a drink in a plastic bottle or use a plastic straw for the next week are things I do on a regular basis to combat feelings of climate doomism. Leveraging feelings of hopelessness and turning them into a stage for activism has really benefited my mental health when trying to deal with these emotions in our fast-moving world. I also feel that in a way, my climate anxiety keeps me motivated to continue in my environmental journey in times when it seems difficult.

Translating Passion to Advocacy

Helen: Grace, in high school, you decided of your own volition to become active in climate policy in a variety of ways. How did you find those opportunities and what drove your decision to apply?
I think one of the main things that drove me to climate activism was seeing how dedicated and inspiring the community is. Discovering there were other youth out there who shared the same ideas, experiences, and passion for environmentalism was both surprising and comforting. When I was younger and had begun to be interested in the climate space, I noticed not a lot of people around me understood what I was thinking about or why it was important. I remember coming home from school one day when I was trying to stop people at school from using plastic water bottles. I was maybe eleven or twelve and saying, "Mom, I feel like no one knows about how important this is! Do these people not know they're killing the turtles?" It was not until later when I became more involved in environmentalism that I realized lots of people do, in fact, care about killing the turtles. Going from a community of people who gave little attention to the environment to seeing people who wanted change as much as I did gave me hope and a reason to keep going and push forward. One of the main ways I found this strength was through online communities and websites. Being able to share the issues happening in my community and finding others who had the same problems in their respective communities showed me I wasn't alone in trying to make a change and made the task of pursuing my passion a lot less daunting. One of the main ways I found these climate spaces was through social media. I applied to my first organization, the Climate Initiative, because I had seen an ad for it when I was scrolling through Instagram. When the opportunity came to apply for a youth ambassador position, I was so excited to hear there was an organization that would back me up on my passion for climate and connect me to a group of people who felt the same. Hearing the ideas shared in this community that welcomed me so openly kept me feeling there was something to fight for, because we were fighting for it together. This is where I feel strongly about using social media and the internet as a tool to inspire activism. Older generations are so quick to discount social media and even blame teens and kids for their "internet addictions," but the truth is I would not be as involved in the

climate space as I am today had it not been for social media. It provides opportunities to participate in new environmental endeavors, it inspires hope by introducing us to a wider network of similarly minded individuals, and it helps push our climate agendas to a larger community than ever before. The simple act of reposting something to your story or replying to a question in your DMs [direct messages] can be the first step in inspiring someone else to get involved. This is what makes social media such a powerful resource: Kids are on social media, so we should be using it to inspire them!

Helen: Grace, why do you think your voice is important in the climate crisis? How have you seen that youth voices matter?
I know my voice is important in the climate crisis because my generation is the one that's going to feel the effects of climate change the most, so it's my responsibility to do something about it. There are plenty of people in older generations who are so passionate about climate change, but at the end of the day, they won't be around to see the planet destroyed in fifty years like we will. They have the passion to make a difference, but not the same urgency as youth have. We know the future of the earth is in our hands, and our future is what we make of it. This is why I feel it's so important to make a change as early as possible.

This feeling was amplified when I attended the Local Conference of Youth (LCOYUSA) in October. The Local Conference of Youth is a youth-led climate conference endorsed by YOUNGO (Youth and Children's Constituency to the United Nations Framework Convention on Climate Change). Every year, the UNFCCC organizes the Conference of Parties, better known as COP. As a precursor to COP, YOUNGO organizes the Conference of Youth (COY), a global event where the Global Youth Statement is formalized and represents young people's demands when presented at COP. Before the COY, local COYs were held on a national level in over seventy countries. During this conference, I witnessed the passion so many young people had for demanding change from their local, state, and national governments, and I saw their dedication to changing the course of global environmental policy for years to come. Hearing people share their experiences of environmental lobbying and climate policymaking gave me a glimpse into what my future could be if I stayed in the climate space. Meeting people at that conference who were congratulating me on being in high school and already beginning my journey in environmentalism made me feel the work I was doing, even if it was just in my small town, was appreciated by more people than just the community it was immediately affecting. It made me feel hopeful that there were such strong voices in the climate movement and helped eliminate my sense of alienation when it came to the climate space. Knowing so many people around me were so dedicated to resolving the same issues I was working on helped me see that the youth voice is more powerful than I had ever seen before. Banding together and fighting for a common goal seemed a lot less daunting than trying to fix a huge problem like climate change by myself and attending the local COY strengthened my hope for the youth movement in a way I had never experienced before.

Grace: Mom, how can you spark initiative in youth voices that aren't directly involved in the climate movement? How do you challenge all students to want to be involved, even when it's not at the forefront of their minds?
So, this is a really hard aspect of my job. I have now experienced environmental education in the K–8 grade band and as a STEM educator, I always have my conservationist hat on. I can't help it: It is where I feel comfortable, it is what I value, and ultimately, it is my life's work. I want all children to understand that our lands, our oceans, and our biodiversity are essential to our existence. While this seems like a basic "duh," when it comes down to teaching this in a classroom, there are a lot of other pressures that take precedence over teaching climate conservation, like teaching kids to read and be mathematically literate! So, my answer to this issue of raising awareness among youth is to connect them to the environment during the school day. If they positively experience land and air in natural spaces there, they will begin to appreciate the need for connection to the environment. Teaching in outdoor spaces helps with this. Learning authentically outside brings delight to student learning and feels different from the "sameness" of every school day. When they are invested in nature, a desire to help in whatever way they can—picking up trash on the soccer field, attending an outdoor environmental camp, connecting with a community conservation group, building bat houses—all those activities provide an entry point to climate awareness and hopefully climate activism. My attitude for teaching kids that don't have a proclivity toward activism is positive praise. Positive recognition of pro-environmental behaviors shows people that no matter how small their contribution is, they will be recognized for it.

Lessons Learned from Our Eco-Relationship

The lessons we've learned as a mother-daughter eco-partnership are many, and we are still discovering them. We have three main takeaways for our readers: First, climate education is a lifelong journey. We want to bring as many people along on our journey as we can, and doing this takes outreach, advocacy, and leading by example. Second, we want people to know that access to nature is essential for healing, motivation, and action. Without spending time in the natural world, our spirits can't renew and connect with others in the shared joy of Mother Nature's splendor. Finally, climate action is not just for one group of people. While it is a responsibility for everyone to shoulder, it is also necessary for everyone to unite in action no matter what age, gender, race, or belief system to which you belong.

Bios

Grace Corveleyn is a high school junior from Pennington, New Jersey. Her environmental passions include conservation, marine science, and climate policymaking. She is currently an Ocean Guardian Ambassador for the National Oceanic and Atmospheric Administration (NOAA), and virtually volunteers with The Climate Initiative. She serves as the co-founder and Vice President of Communications of the Youth Oceanic Initiative, a non-profit organization dedicated to making the environmental movement more accessible by spreading the messaging of ocean conservation to students from around the world. She attended the Local Conference of Youth for the United States (LCOYUSA) in 2023, drafting the national youth statement of climate demands that was presented at the 28th Conference of Parties (COP28) last December. In the future, Grace hopes to continue her passion for marine and environmental science through her college education and aspires to pursue a career in marine biology. In addition to her work in the climate space, Grace is a three-sport athlete, avid Girl Scout, and loves spending time outside with her friends and family.

Helen Corveleyn has experience in PreK–12 STEM education. She is a STEM teacher at Timberlane Middle School and spent ten years as a STEM facilitator at Hopewell Elementary School. She has a M.A.T. in conservation biology from Miami University of Ohio, which complements her B.S. in environmental policy from Marist College. Helen is the recipient of the Presidential Award in Math and Science Teaching given by the National Science Foundation and the White House. Other honors include receiving the Governor's Educator of the Year Award (2019), the New Jersey County Teacher of the Year (2020), and the "I Can STEM NJ Role Model" for the New Jersey STEM Pathways Network (2021). Most recently, she received the Mercer County Society of Professional Engineers STEM Teacher of the Year Award (2024). Her passion is inspiring young people to become planetary stewards who can communicate scientific ideas and promote innovation in science and sustainability. She teaches graduate students in environmental leadership at The College of New Jersey and in conservation biology at Miami University of Ohio. Her international fieldwork includes studying island biogeography and swimming with whale sharks in Baja, Mexico; studying orangutans and sustainable palm oil in Borneo, Malaysia; and creating a multimedia-based conservation campaign to support the Belize Zoo and Maya Forest Corridor.

Further Reading

Miri Yemini, Laura Engel, and Adi Ben Simon (2023) Place-based education – a systematic review of literature, Educational Review, DOI: 10.1080/00131911.2023.21772Works Cited

Taylor & Francis Online." Accessed November 9, 2023. https://www.tandfonline.com/action/showCitFormats?doi=10.1080%2F00131911.2023.2177260

"Marking the Kyoto Protocol's 25th anniversary | United Nations." United Nations. Accessed November 11, 2023. https://www.un.org/en/climatechange/marking-kyoto-protocol%E2%80%99s-25th-anniversary.

CHAPTER 20

Effective Climate Change Education in the Classroom Begins with Effective Climate Change Professional Learning for Educators

Karen Woodruff, Missy Holzer, and Andrea Drewes

Teacher professional learning (PL) focused on climate change is crucial for equipping educators with the knowledge and skills necessary to effectively teach about this pressing global issue. Educators are on the frontline in the world of climate change education, teaching students the interdisciplinary content necessary to understand the complexities and associated impacts of climate change with the goal of inspiring the generation of viable solutions for mitigating human-induced changes. To be effective in this role, educators need a solid background in the causes and effects of climate change and effective practices for teaching students across all grade bands and subject areas.

Climate change PL models currently exist in many forms, with the intention of supporting educators in developing the pedagogical content knowledge (PCK) they need to integrate climate change in meaningful ways. The wide variety of PL design and implementation models that exist across contexts provides an opportunity to review the elements that should be considered when creating effective future PL programs for climate change education. This chapter offers a comprehensive framework for the design of effective climate change PL that leverages components of exemplary programs and serves as a guideline for the development of future PL programs that center students and the places they learn.

Core Components of Teacher Professional Learning in the Literature

PL on climate change exists in a variety of durations, locations, modalities, and foci on climate change–related topics. However, common design elements across programs have a positive influence on teacher learning.[1] To situate this chapter in current best practices within the field and make recommendations for the design of effective PL, we begin with a targeted review of the literature on climate change–focused PL. Our review suggests that effective PL design includes attention to teacher content knowledge and pedagogy, alignment with educational theory and professional standards for climate change, and consideration of PL objectives and local teacher and student priorities.

Attention to both teacher content knowledge and pedagogy is a fundamental component of effective PL.[2] When teachers lack confidence with content, they can be reluctant to teach it.[3,4] Likewise, comfort with content alone does not imply teachers use the practices that are best suited for teaching students.[5] Teaching climate change can suffer from both lack of content knowledge and lack of pedagogical skills, suggesting there is a need to support teachers in overcoming these challenges through effective PL models that attend to PCK.[6] Teachers may wish to deepen their understanding about climate change but lack formal support from their school district to do so.[7] To determine where teachers do find support, Puttick and Talks[8] conducted a scoping review and revealed that teachers use government websites such as NASA and NOAA and mass media as primary sources of information. They also rely on external teacher PL, as most districts do not dedicate district PL time to climate change.[9]

Nation and Feldman[10] offered and studied one external PL opportunity in the southeastern United States that engaged marine science teachers in PL activities monthly for eight months during the school year. The data suggested participating teachers' knowledge of climate shifted over time and the teachers who had originally shared concerns about teaching climate change gained confidence with these topics. Teachers reported feeling pressure to remain neutral on the topic at the beginning of the PL; however, after having time to understand the topics and their connections to interdisciplinary content, they reported no longer feeling such pressure. The collaborative nature of the PL design and teachers' ability to practice strategies with colleagues helped them feel confident enough to teach climate change.

Similarly, in the suburban Midwest, teacher educators implemented a weeklong workshop with nineteen secondary science teachers from schools serving Native American students. The PL occurred in collaboration with NASA's Innovations in Climate Education project, with the goal of enhancing climate change literacy in Native American communities. Using an argumentation approach, teachers had time to make sense of evidence of human-induced climate change. As a result, they demonstrated shifts in their beliefs about climate change over time. The blend of content and pedagogy provided teachers with opportunities to reason with content while learning practices to effectively teach their students using appropriate methods for their specific grade levels and contexts.[11]

The literature also suggests teachers are more likely to engage in PL when they understand the reason for the program design and how it aligns with the disciplinary standards by which they are evaluated. Teaching climate change is interdisciplinary, with applications across content areas. A full understanding of climate change's causes, effects, impacts, and mitigation efforts requires knowledge in all traditional disciplines taught in schools. Programs that embed climate change in the context of disciplinary content provide a theoretical basis for teaching climate change. Given the constructivist approach that is central to state and national standards and curricula, teaching about climate change offers opportunities for application and synthesis of knowledge. Furthermore, interest in climate change often comes from students, providing an entry point for culturally embedded real-world application of concepts. When well-integrated with the foundational theories of teaching and learning, PL programs can depoliticize climate change, focusing on learning and student engagement in sensemaking about core ideas in which they are interested.

Finally, there is sufficient evidence to suggest that alignment (or misalignment) with local priorities is a key component of PL design. While recent studies suggest teachers are concerned about climate change and want to teach it, they often lack support to do so. Understanding the causes of misalignment is necessary to address change. For example, when Ennes et al.[12] sought to understand what prevented teachers from participating in climate change focused PL, they discovered most teachers believed climate change was occurring. They started the PL program thinking politically grounded ideas would be the greatest barrier to teaching climate change but discovered that politics was not the most significant concern. The greatest barrier was time and lack of clarity about connections to the content they taught. Based on this finding, the authors suggested embedding PL within the school setting and demonstrating integration of climate topics with standards-based content.

Alignment can also include drawing connections between local contexts and the theoretical reasoning behind the work. The West Virginia Climate Change Professional Development Project, a collaborative effort between social scientists, educators, and local stakeholders in climate change, was designed to increase collaboration and empower teachers to use data-driven investigations in the classroom. Program designers recognized the local connections in a fossil fuel–reliant state and focused on supporting teachers in developing the knowledge and confidence to teach about the complexities of energy development.[13] Similarly, Bloom and Quebec Fuentes[14] addressed energy production using an experiential learning approach that integrated mathematical modeling to provide teachers the tools necessary to challenge subjective perspectives with evidence. Both approaches found that teachers increased their confidence in teaching about energy and climate change, and most shifted their perceptions about climate change overall. The practice of grounding PL in local contexts allowed teachers to connect with the theoretical reasoning behind the work.

The North American Association for Environmental Education reports that teachers who believe their district and state support teaching climate change are more likely to feel prepared and to teach the topic. The programs reviewed also suggest that alignment between PL objectives and the school setting is essential for PL to be effective. Teachers may individually engage in innovative teaching practices; however, within the field of professional development, there is agreement that broad change is not possible without a coherent understanding of its purpose and a collaborative environment for problem-solving.[15] While models for climate change PL vary, the common elements of attention to PCK, alignment with professional standards, and coherence and collaboration with schoolwide priorities are consistent components of successful PL programs. Leveraging this knowledge, we recommend a set of specific foundational elements of PL and design components to consider when developing models for climate change-focused PL.

A Professional Learning Design Framework for Climate Change

Designing effective PL programs is akin to building a customized home: Start with a solid foundation and select from a variety of options to create the most optimal design for the context. We use this analogy to suggest the professional learning design framework (figure 20.1). A solid foundation and strategic selection of design components supports the needs of teachers and students in specific contexts, ensuring the program's success. The professional learning design framework recommends that PL foundations be considered in all designs and that design components be consistently revisited throughout the iterative process.

PROFESSIONAL LEARNING FOUNDATIONS

The PL foundations ensure the PL includes coherent, collaboratively developed goals; alignment with professional standards; and the support of administrators and other stakeholders.

Coherent, Collaboratively Developed Goals

PL programs should be designed collaboratively with individuals focused on the PL's overarching goals within the teaching and learning context. A clear plan

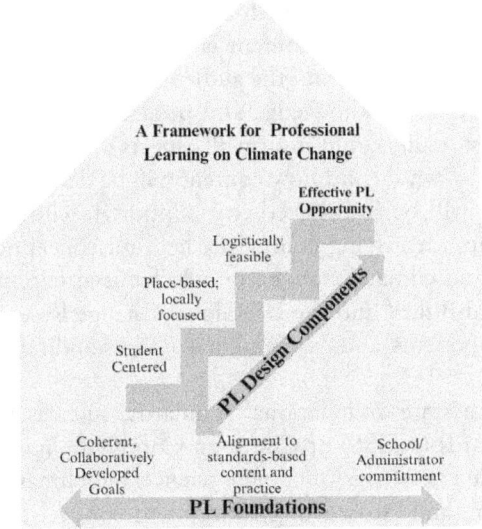

Figure 20.1 Framework for designing effective professional learning opportunities on climate change

for the work within a given time period, stakeholders' specific responsibilities, and time for reflection and redirection will ensure a collaborative program that is responsive to stakeholder needs. A design-based approach can ensure the plan is iterative and allows for teacher voice and agency in the process.[16] Individuals designing the PL may ask questions like the following: Is there a funding partner with an agenda that needs to be considered? What are the desired outcomes for the PL? How will the context (presenters, audience, hosts, etc.) for the PL program be factored into achieving the goals? How will the identified goals support student learning about climate change? How will researchers measure student learning? What are the school district's and community's local assets? How will we measure the PL program's effect? If the PL program is intended to be regional or national in its reach, consider including climate change topics that will have meaning to as many participants as possible. These questions and others will help narrow the PL program's goals through collaborative discussion.

Alignment with Standards-Based Content and Practice

Designing PL programs includes identifying participating teachers' learning needs regarding specific disciplinary content, how people learn specific disciplinary content, and how to teach specific disciplinary content. Know the audience prior to the design of the PL so the appropriate type and amount of content is integrated.[17] For example, if the audience includes only science educators, it is important to identify their level of understanding of climate change content relative to the grades they teach and to identify the science topics they teach to ensure the PL program's content is relevant to their teaching contexts. One method of gathering data from the audience is to survey attending teachers about their specific content knowledge and pedagogical challenges in teaching that content. Survey results could inform all aspects of the PL program's design, including logistics, by whom and how content will be delivered, and what pedagogical approaches will be introduced that support teaching and learning the content. In addition, identifying commonly held misconceptions about aspects of climate science and climate change can help focus attention on specific content. McCaffrey and Buhr[18] and the US Global Change Research Program have identified key components and fundamental understandings of major climate change topics.

Alignment with state and national standards, such as the Next Generation Science Standards (NGSS) or Common Core State Standards, is essential. NGSS includes core knowledge of climate science and climate change, practices to understand them, and crosscutting concepts to connect climate science and climate change across science content areas in all grade levels. The core ideas of climate change, such as those recommended by Windschitl,[19] may serve as guidelines. Windschitl[20] identified five key topic areas that should serve as foundational knowledge when unpacking the complexities of our climate system and climate change, including the greenhouse effect, the carbon cycle, oceans and cryosphere, ecosystems and biodiversity, and extreme weather.[21] Breslyn et al.[22] expand the progression of learning climate change to include the role of human activity, the greenhouse effect mechanism, impacts of climate change, and mitigation and adaptation strategies. We recommend reviewing existing literature, standards, and guidelines from reputable sources to frame this effort.

Administrative Support and Commitment

School administrators' support and commitment are instrumental to successfully implementing climate change education initiatives and PL programs, as administrators play a critical role in enabling educators to enhance their teaching practices effectively. Lack of support from principals, other administrators, and/or peers is the primary reason initiatives fail to bring about change in teaching practices.[23] By advocating for climate-change education within the school community, administrators demonstrate a commitment to sustainability, environmental awareness, and social responsibility, shaping a culture of continuous learning and professional growth. Administrator engagement can create an environment where educators feel empowered to integrate climate change education into their instructional practices and drive positive change within their schools. Consider engaging administration in the development process.

PL DESIGN COMPONENTS

Once the design's foundation is established, it is necessary to consider specific components of the PL. We suggest that the components be grounded in students' needs and interests; that they center local places, concerns, and assets; and that the design team consider the logistics of implementation.

Student-Centered

Establishing a collective vision for why the PL experience is worth teachers' time, effort, and dedication requires getting to know the stakeholders involved and coalescing around a shared purpose. Students are the driving reason for developing PL for teachers. Grounding the PL program's focus in contexts meaningful to students aligns with current research-based approaches to how students learn best. During the first component in the design process, consider who the students are, issues important in their local communities, and the cultural assets they bring to the learning environment. This work requires getting to know students and what is important to them.

Being responsive to student interests provides an opportunity to engage local community groups, organizations, and industry in school communities. These groups can be assets in PL. For example, several national organizations are designing innovative technologies and solutions to mitigate the effects of climate change and remove carbon dioxide from the atmosphere, and nearby school districts may benefit from forging relationships with these organizations.[24] Local community groups and nonprofits work with interconnected systems related to climate change, including oceans, ecosystems, weather, and more. Identifying the professionals working in these spaces, partnering with them, and connecting their work to the specific content addressed in the PL experience can enhance the effort's effectiveness. Educators can invite professionals into the classroom or bring students to work sites if appropriate. They often provide an experiential opportunity to engage in aspects of climate-focused efforts and provide hands-on experiences. Examples include opportunities in energy production,[25] policy advocacy, tree-planting, community gardening,[26] or school-based greening initiatives (see Chapter 8 in this book). Industry engagement also allows students to apply the concepts educators teach and emphasizes the skills necessary for students' future careers. When driven by student interests, community partnerships can be successful enhancements to PL program design.

Place-Based and Locally Focused

Place-based education models offer a powerful approach to integrating environmental content into the curriculum while deepening students' connection with their local environment. The richness of local communities, cultures, and landscapes is woven into learning opportunities through "place-based" pedagogy.[27] This approach embeds learning in local contexts and invites community members to collaborate with educators and students. While related to the community partnerships discussed above, place-based learning includes hands-on experiential opportunities that allow learners (teachers in PL, or students in classrooms) to explore and understand local ecosystems, analyze environmental issues, and develop contextually relevant solutions. This approach allows educators to promote environmental literacy, critical thinking skills, and a strong sense of stewardship. Incorporating place-based education models into teacher PL can equip educators with the tools to foster environmental awareness and inspire action in their classrooms.

An exemplary project, "Learning in Places: PK-5+ Field-Based Science Education Across Schools, Families, and Communities,"[28] provides guidance for designing outdoor learning sites and collaborating with community members to enhance learning. The project engages elementary school children in learning about food security by establishing sustainable gardens in partnership with the Tilth Alliance, a local agricultural organization in Seattle, Washington. Place-based education centers active engagement in service projects within the local school and community as a key aspect of learning and often integrates across disciplinary content areas, including language arts, mathematics, social studies, science, and various other subjects throughout the curriculum.

Place-based education also provides opportunities to center students' ideas and what is important to them in their everyday lives. The practices teachers employ in the classroom are essential to student success and can result in students feeling either welcomed or excluded in classrooms. Pedagogies that center students' ideas and value what they already know about a topic tend to be inclusionary. These ways of facilitating classroom teaching center on how students think about ideas and build understanding from where they are to where they need to be to demonstrate subject mastery.[29] PL designed to center what students know about climate change from personal experience, the media, and those who influence them provides students a voice in the learning process. Learning to facilitate classrooms where students have voice and agency and where their relevant experiences are valued is essential for effective climate change PL.

Logistically Feasible

Effective logistical planning is essential for designing successful climate change PL. Securing adequate funding for activities is crucial to ensure educators have access to high-quality resources, training materials, and external expertise to enhance their knowledge and skills. Additionally, allocating sufficient time for PL allows educators to engage in meaningful discussions, interactive activities, and collaborative projects that deepen their understanding of climate change concepts and teaching strategies. The lack of dedicated time to engage in climate change PL is a primary barrier for many teachers.[30] Duration is a critical consideration in developing PL, as there is near-universal agreement that "one-and-done" workshop models are rarely successful[31] unless they are designed with a very specific purpose. Instead, models that allow teachers to repeatedly gather over time, such as an academic year or longer, report greater effectiveness.[32] The appropriate time in the context of other initiatives, the venue or platform, materials, resources, budget, scheduling, travel, accessibility, and inclusivity of the program should all be collaboratively discussed with the design team. Once planning decisions are made, a final review must be made to evaluate the PL program's coherence for continuity between component design choices and the program's foundational elements.

THE FRAMEWORK IN ACTION

The framework draws upon recognized best practices in the field of climate change PL, drawing attention to foundational aspects of effective PL design and components that align the PL with students and their communities. The programs presented below serve as examples of PL that exemplify the integration of these key aspects of the framework.

PRISM Center Professional Development

The Montclair State University PRISM Center hosted a year-long PL opportunity for educators interested in integrating climate change into their classrooms. Participants met six times over the course of the school year and interacted online between in-person sessions. Beginning with an overview of the context of climate change education in New Jersey, the group explored evidence for climate change through various datasets organized through the lens of the earth's spheres: atmosphere, hydrosphere, cryosphere, geosphere, and biosphere. The interdisciplinary content knowledge necessary to understand changes taking place within each sphere was the focal point for each session. Hands-on lab activities, datasets, and opportunities to anticipate student questions were incorporated into each of the six meetings.

Participants' responses to this format were positive. Specifically, educators reported increased confidence, a deeper understanding of global climate-change impacts, and an enthusiasm for discussing climate issues. The workshop provided a wealth of resources, fostering a sense of preparedness and inspiration for integrating climate change topics into participants' teaching. They valued the abundance of data and resources, including charts for inquiry-based lessons, which empowered teachers with valuable tools for effective and engaging climate science instruction. Teachers created lessons within their groups and appreciated the exchange of practical strategies for communicating climate-change topics and the opportunity to interact with fellow educators. The design and implementation of this in-person workshop over a one-year timeframe allowed teachers to come together in person, make sense of ideas and resources, and return to their classrooms to work with students in between sessions.

Endeavor STEM Teaching Certificate Project Online Graduate Course

Online learning can be a viable modality for working across geographically distant contexts. The Endeavor STEM Teaching Certificate Project, a partner with NASA's Office of STEM Engagement (Space Act Agreement #18129), offers a graduate course called "Climate Science: Socioscientific Issues in the STEM Classroom." Educators learn the interdisciplinary nature of climate science and connections to content they teach in their NGSS-based classrooms. Using NASA climate resources, educators focus on developing activities that provide students with data analysis experience, critical thinking, and argumentation skills within their specific teaching contexts. Educators report increased awareness of and confidence in teaching climate change and emphasize that the NASA data provide evidence to support data-driven conversations in their classrooms. The course's online venue provides opportunities for educators across the United Stated to share lessons, activities, resources, data, and demonstrations. Upon completing the program, educators turnkey their development through the design of a PL opportunity for their colleagues, aligned with the framework's foundational elements and components.

MADE CLEAR Climate Academies

The MADE CLEAR project was implemented over ten years beginning in 2013 and serves more than three hundred educators, primarily in Maryland and Delaware, as well as throughout the Mid-Atlantic region through yearly Climate Academies. Each Climate Academy was designed to provide current climate science content, model lessons and learning activities, and practical experiences investigating the socio-scientific issue of climate change and supporting educators to bring the topic to their students. The Climate Academy's format focused on an in-person, weeklong residential program at an environmental education center with four virtual follow-up sessions and two full-day in-person sessions spread across the academic year to promote sustained engagement. Teacher participation was incentivized through financial support from a National Science Foundation grant. A cross-disciplinary design team of climate researchers, learning scientists, teacher educators, and practitioners developed the PL experience to deepen teachers' climate change content knowledge, while providing pedagogical support and instructional resources needed to engage in climate education tailored to each teacher's unique context.

The MADE CLEAR Climate Academy specifically addressed content found to challenge teachers and students alike. Connections to state and national standards can help align climate change with teacher priorities.[33] Early academies unintentionally focused more on the science and evidence-driven information about the greenhouse mechanism and climate change effects and less on solutions, such as mitigation and adaptation strategies, and socio-scientific and cross-disciplinary connections. Facilitators addressed this unbalanced approach through research into the conceptual mobility of climate change topics.[34] The team identified appropriate teaching resources and technologies that supported NGSS and grounded the PL in research-based practices. Subsequent Climate Academies also varied in geographic location, session logistics, content focus, and interdisciplinary design but continues today under the efforts of the Mid-Atlantic Climate Change Education Collaborative. Analysis of Climate Academy attendees' reflections indicated that they found the content beneficial for providing content related to climate change aligned with their standards.[35]

Climate on My Campus: Using Field-Based Data in the Classroom

In some situations, resources are available to offer a "one-and-done"-style PL opportunity. Although they are not the ideal design for supporting teachers in climate change education, there are examples of programs that identify and plan for a specific goal. Climate on My Campus was implemented using funding allocated for a one-day PL event that connected with the funder's mission. Two educators planned the event at an outdoor environmental education campus. A scientist who used the environmental education campus as a living laboratory shared his expertise and taught participants how to collect and analyze tree ring data. The PL event included informal and formal education connections through a place-based, student-centered learning opportunity. Content selection included teaching resources readily available and replicable for participants. The teaching methods were inquiry-focused and required participants to act as students except for during reflection periods. Finally, the event's logistical aspects were identified throughout the entire planning process and were supported by the environmental education center's team. Only a few activities were presented because they could be easily replicated at the home campuses of educators in attendance. To ensure the connection to the one-day event was sustained, PL leaders made themselves available beyond the workshop, and participants all stated they would be interested in participating in a series of climate change workshops in the future. The success of this event was due in part to the connection between formal and informal educators in the program's design and implementation. This connection brought together perspectives on effective teaching and learning and perspectives on how to engage learners with their local environment. The synergy between informal and formal educators offers possibilities for meaningful locally focused PL opportunities.

FORWARD THINKING: DESIGNING EFFECTIVE PL FOR CLIMATE CHANGE

When designing PL to support teachers' climate change-related knowledge and pedagogy, foundations must be put in place before selecting design components. The professional learning design framework draws on best practices in the field of PL, considers the specific aspects of climate change content and pedagogy, and includes flexibility and responsiveness. When using the framework as your guide, recognize you may encounter context-specific roadblocks along the way. The framework's components are intended to allow for flexibility within local contexts, giving voice to stakeholders engaged in the work. Taking the time to collaboratively identify potential roadblocks for all aspects of the PL program design is the best way to ensure viable solutions are in place to prevent barriers to supporting teachers with the PL they need through the program's implementation. For example, if administrative support is lacking, consider developing a list of ways the PL program will manifest itself in positive ways within the district and share that list with administrators. If funding is preventing educators from participating, consider providing them with a list of funding streams, such as local organizations or education foundations. If possible, secure funding prior to engaging in PL. Plan with locally mandated curriculum in mind and take the time to identify how the PL program goals could directly connect to local curricula.

Effective climate-change education begins with effective climate change PL for educators. We invite you to use the PL design framework to design a PL program that supports educators on their climate change education journeys.

Notes

1. Mary M. Kennedy, "How Does Professional Development Improve Teaching?," *Review of Educational Research* 86, no. 4 (2016), doi:10.3102/0034654315626800.

2. Laura M. Desimone et al., "Formal and Informal Mentoring: Complementary, Compensatory, or Consistent?," *Journal of Teacher Education* 65, no. 2 (2014), doi:10.1177/0022487113511643.

3. Mark Windschitl and David Stroupe, *The Three-Story Challenge: Implications of the Next Generation Science Standards for Teacher Preparation* 68, no. 3 (2017), doi:10.1177/0022487117696278

4. Mark Windschitl, "Inquiry Projects in Science Teacher Education: What Can Investigative Experiences Reveal About Teacher Thinking and Eventual Classroom Practice?," *Science Education* 87, no. 1 (2003), doi:10.1002/sce.10044

5. Okhee Lee et al., "Professional Development in Inquiry-Based Science for Elementary Teachers of Diverse Student Groups," *Journal of Research in Science Teaching* 41, no. 10 (2004), doi:10.1002/tea.20037

6. Megan Ennes et al., "It's About Time: Perceived Barriers to In-Service Teacher Climate Change Professional Development," *Environmental Education Research* 27, no. 5 (2021), doi:10.1080/13504622.2021.1909708

7. Steven Puttick and Isobel Talks, "Teachers' Sources of Information about Climate Change: A Scoping Review," *The Curriculum Journal* 33, no. 3 (2022): 378–95.

8. Puttick & Talks, "Teachers' Sources of Information."

9. Andrea Drewes, Joseph Henderson, and Chrystalla Mouza, "Professional Development Design Considerations in Climate Change Education: Teacher Enactment and Student Learning," *International Journal of Science Education* 40, no. 1 (2018), doi:10.1080/09500693.2017.1397798; Puttick & Talks, "Teachers' Sources of Information."

10. Molly Trendell Nation and Allan Feldman, "Climate Change and Political Controversy in the Science Classroom: How Teachers' Beliefs Influence Instruction," *Science & Education* 31, no. 6 (2021), doi:10.1007/s11191-022-00330-6.

11. Shiyu Liu and Gillian Roehrig, "Exploring Science Teachers' Argumentation and Personal Epistemology About Global Climate Change," *Research in Science Education* 49, no. 1 (2019): 173-189; Gillian Roehrig et al., "CYCLES: A Culturally-Relevant Approach to Climate Change Education in Native Communities," *Journal of Curriculum & Instruction* 6, no. 1 (2012): 73–89, doi:10.3776/joci.2012.v6n1p73-89

12. Ennes et al., "It's About Time."

13. Kathryn Williamson et al., "A Case Study for Climate Change Teacher Professional Development in West Virginia," *Journal of Sustainability Education* 28; North American Association for Environmental Education, *2022 Annual Report* (Washington, DC: NAAEE, 2022).

14. Mark Bloom and Sarah Quebec Fuentes, "Experiential Learning for Enhancing Environmental Literacy Regarding Energy: A Professional Development Program for Inservice Science Teachers," *Eurasia Journal of Mathematics, Science and Technology Education* 15, no. 6 (2019): em1699, doi:10.29333/ejmste/103571.

15. Anthony S. Bryk et al., *Learning to Improve: How America's Schools Can Get Better at Getting Better* (Cambridge, MA: Harvard Education Press, 2015); William R. Penuel et al., "What Makes Professional Development Effective? Strategies That Foster Curriculum Implementation," *American Educational Research Journal* 44, no. 4 (2007), doi:10.3102/0002831207308221; Jonathan A. Supovitz and Herbert M. Turner, "The Effects of Professional Development on Science Teaching Practices and Classroom Culture," *Journal of Research in Science Teaching* 37, no. 9 (2000), doi:10.1002/1098-2736(200011)37:9<963::AID-TEA6>3.0CO;2-0.

16. William R. Penuel and Ashley Seidel Potvin, "Design-Based Implementation Research to Support Inquiry Learning," in *International Handbook of Inquiry and Learning*, ed. Ravit Golan Duncan and Clark Chinn (London, UK: Routledge, 2021).

17. Susan Loucks-Horsley et al., *Designing Professional Development for Teachers of Science and Mathematics* (Thousand Oaks, CA: Corwin Press, 2009).

18. Mark S. McCaffrey and Susan M. Buhr, "Clarifying Climate Confusion: Addressing Systemic Holes, Cognitive Gaps, and Misconceptions through Climate Literacy," *Physical Geography* 29, no. 6 (2008): 512–28, doi:10.2747/0272-3646.29.6.512.

19. Mark Windschitl, *Teaching Climate Change: Fostering Understanding, Resilience, and a Commitment to Justice* (Cambridge, MA: Harvard Education Press, 2023).

20. Windschitl, *Teaching Climate Change*.

21. Wayne Breslyn and J Randy McGinnis, "Investigating Preservice Elementary Science Teachers' Understanding of Climate Change from a Computational Thinking Systems Perspective," *Eurasia Journal of Mathematics, Science and Technology Education* 15, no. 6 (2019): em1696, doi:10.29333/ejmste/103566.

22. Wayne Breslyn et al., "Development of an Empirically-Based Conditional Learning Progression for Climate Change," *Science Education International* 28, no. 3 (2017): 214–23.

23. Karen S. Acton, "Environmental Teacher Leadership: Overcoming Barriers Posed by School Culture, School Structure, and the Principal," *International Journal of Leadership in Education* (2022): 1–21, doi:10.1080/13603124.2022.2032369. Joan L. Buttram and Elizabeth N. Farley-Ripple, "The Role of Principals in Professional Learning Communities," *Leadership and Policy in Schools* 15, no. 2 (2016): 192-220, doi:10.1080/15700763.2015.1039136; Minna Körkkö et al., "Developing Teacher In-Service Education through a Professional Development Plan: Modelling the Process," *European Journal of Teacher Education* 45, no. 3 (2022): 320–37, doi:10.1080/02619768.2020.1827393.

24. Project Drawdown, https://drawdown.org/drawdown-foundations.

25. Bloom and Quebec Fuentes, "Experiential Learning."

26. Carlie D. Trott, "Reshaping Our World: Collaborating with Children for Community-Based Climate Change Action," *Action Research* 17, no. 1 (2019): 42–62, doi:10.1177/1476750319829209.

27. Robert Barratt and Elisabeth Barratt Hacking, "Place-Based Education and Practice: Observations from the Field," *Children Youth and Environments* 21, no. 1 (2011): 1–13, doi:10.7721/chilyoutenvi.21.1.0001; Yi Sun, Roger C. K. Chan, and Huiwei Chen, "Learning with Geographical Sensitivity: Place-Based Education and Its Praxis," *The Professional Geographer* 68, no. 4 (2016): 574–583, doi:10.1080/00330124.2015.1121835.

28. Megan Bang, "Learning on the Move Toward Just, Sustainable, and Culturally Thriving Futures," *Cognition and Instruction* 38, no. 3 (2020): 434–44, doi:10.1080/07370008.2020.1777999.

29. Mark Windschitl, Jessica Thompson, and Melissa Braaten, *Ambitious Science Teaching* (Cambridge, MA: Harvard Education Press, 2018).

30. Ennes et al., "It's About Time."

31. Erin Redman, Arnim Wiek, and Aaron Redman, "Continuing Professional Development in Sustainability Education for K–12 Teachers: Principles, Programme, Ap-

plications, Outlook," *Journal of Education for Sustainable Development* 12, no. 1 (2018), doi:10.1177/2455133318777182; Penuel et al., "What Makes Professional Development Effective?"

32. Nicole Shea, Chrystalla Mouza, and Andrea Drewes, "Climate Change Professional Development: Design, Implementation, and Initial Outcomes on Teacher Learning, Practice, and Student Beliefs," *Journal of Science Teacher Education* 27, no. 3 (2016): 235–258; Michael S. Garet et al., "What Makes Professional Development Effective? Results from a National Sample of Teachers," *American Educational Research Journal* 38, no. 4 (2001), doi:10.3102/00028312038004915.

33. Drewes et al., "Professional Development Design"; Emily Hestness et al., "Science Teacher Professional Development in Climate Change Education Informed by the Next Generation Science Standards," *Journal of Geoscience Education* 62, no. 3 (2014): 319–329, doi:10.5408/13-049.1; Emily Hestness et al., "Examining Science Educators' Perspectives on Learning Progressions in a Climate Change Education Professional Development Program," *Journal of Science Teacher Education* 28, no. 3 (2017): 250–74, doi:10.1080/1046560X.2017.1302728.

34. Drewes et al., "Professional Development Design."

35. Nicole A. Shea, Chrystalla Mouza, and Andrea Drewes, "Climate Change Professional Development: Design, Implementation, and Initial Outcomes on Teacher Learning, Practice, and Student Beliefs," *Journal of Science Teacher Education* 27, no. 3 (2016): 235–58.

References

Acton, Karen S. "Environmental Teacher Leadership: Overcoming Barriers Posed by School Culture, School Structure, and the Principal." *International Journal of Leadership in Education* (2022): 1–21.

Bang, Megan. "Learning on the Move toward Just, Sustainable, and Culturally Thriving Futures." *Cognition and Instruction* 38, no. 3 (2020): 434–44.

Barratt, Robert, and Elisabeth Barratt Hacking. "Place-Based Education and Practice: Observations from the Field." *Children Youth and Environments* 21, no. 1 (2011): 1–13.

Bloom, Mark, and Sarah Quebec Fuentes. "Experiential Learning for Enhancing Environmental Literacy Regarding Energy: A Professional Development Program for Inservice Science Teachers." (2019).

Breslyn, Wayne, Andrea Drewes, J. Randy McGinnis, Emily Hestness, and Chrystalla Mouza. "Development of an Empirically-Based Conditional Learning Progression for Climate Change." *Science Education International* 28, no. 3 (2017).

Breslyn, Wayne, and J Randy McGinnis. "Investigating Preservice Elementary Science Teachers' Understanding of Climate Change from a Computational Thinking Systems Perspective." *Eurasia Journal of Mathematics, Science and Technology Education* 15, no. 6 (2019): em1696.

Bryk, Anthony S., Louis M. Gomez, Alicia Grunow, and Paul G. LeMahieu. *Learning to Improve: How America's Schools Can Get Better at Getting Better*. Cambridge, MA: Harvard Education Press, 2015.

Buttram, Joan L., and Elizabeth N. Farley-Ripple. "The Role of Principals in Professional Learning Communities." *Leadership and Policy in Schools* 15, no. 2 (2016): 192–220.

Desimone, L. M., E. D. Hochberg, A. C. Porter, M. S. Polikoff, R. Schwartz, and L. J. Johnson. "Formal and Informal Mentoring: Complementary, Compensatory, or Consistent?" *Journal of Teacher Education* 65, no. 2 (Mar 2014): 88–110.

Drewes, Andrea, Joseph Henderson, and Chrystalla Mouza. "Professional Development Design Considerations in Climate Change Education: Teacher Enactment and Student Learning." *International Journal of Science Education* 40, no. 1 (2018): 67–89.

Education, North American Association for Environmental. "2022 Annual Report." (2022).

Ennes, Megan, Danielle F. Lawson, Kathryn T. Stevenson, M. Nils Peterson, and M. Gail Jones. "It's About Time: Perceived Barriers to in-Service Teacher Climate Change Professional Development." *Environmental Education Research* 27, no. 5 (2021): 762–78.

Garet, Michael S., Andrew C. Porter, Laura Desimone, Beatrice F. Birman, and Kwang Suk Yoon. "What Makes Professional Development Effective? Results from a National Sample of Teachers." *American educational research journal* 38, no. 4 (2001): 915–45.

Hestness, Emily, R. Christopher McDonald, Wayne Breslyn, J. Randy McGinnis, and Chrystalla Mouza. "Science Teacher Professional Development in Climate Change Education Informed by the Next Generation Science Standards." *Journal of Geoscience Education* 62, no. 3 (2014): 319–29.

Hestness, Emily, J. Randy McGinnis, Wayne Breslyn, R. Christopher McDonald, and Chrystalla Mouza. "Examining Science Educators' Perspectives on Learning Progressions in a Climate Change Education Professional Development Program." *Journal of Science Teacher Education* 28, no. 3 (2017): 250–74.

Kennedy, Mary M. "How Does Professional Development Improve Teaching?" *Review of educational research* 86, no. 4 (2016): 945–80.

Körkkö, Minna, Marja-Riitta Kotilainen, Sanna Toljamo, and Tuija Turunen. "Developing Teacher in-Service Education through a Professional Development Plan: Modelling the Process." *European Journal of Teacher Education* 45, no. 3 (2022): 320–37.

Lee, Okhee, Juliet E. Hart, Peggy Cuevas, and Craig Enders. "Professional Development in Inquiry-Based Science for Elementary Teachers of Diverse Student Groups." *Journal of research in science teaching* 41, no. 10 (2004): 1021–43.

Liu, Shiyu, and Gillian Roehrig. "Exploring Science Teachers' Argumentation and Personal Epistemology About Global Climate Change." *Research in Science Education* 49, no. 1 (2019): 173–89.

Loucks-Horsley, Susan, Katherine E. Stiles, Susan Mundry, Nancy Love, and Peter W. Hewson. *Designing Professional Development for Teachers of Science and Mathematics*. Corwin press, 2009.

McCaffrey, Mark S., and Susan M. Buhr. "Clarifying Climate Confusion: Addressing Systemic Holes, Cognitive Gaps, and Misconceptions through Climate Literacy." *Physical Geography* 29, no. 6 (2008): 512–28.

Nation, Molly Trendell, and Allan Feldman. "Climate Change and Political Controversy in the Science Classroom: How Teachers' Beliefs Influence Instruction." *Science & education* 31, no. 6 (2022): 1567-83.

Penuel, William R., Barry J. Fishman, Ryoko Yamaguchi, and Lawrence P. Gallagher. "What Makes Professional Development Effective? Strategies That Foster Curriculum Implementation." *American Educational Research Journal* 44, no. 4 (2007): 921–58.

Penuel, William R., and Ashley Seidel Potvin. "Design-Based Implementation Research to Support Inquiry Learning." In *International Handbook of Inquiry and Learning*, 74–87: Routledge, 2021.

Project_Drawdown. https://drawdown.org/drawdown-foundations.

Puttick, Steven, and Isobel Talks. "Teachers' Sources of Information About Climate Change: A Scoping Review." *The Curriculum Journal* 33, no. 3 (2022): 378–95.

Redman, Erin, Arnim Wiek, and Aaron Redman. "Continuing Professional Development in Sustainability Education for K-12 Teachers: Principles, Programme, Applications, Outlook." *Journal of Education for Sustainable Development* 12, no. 1 (2018): 59–80.

Roehrig, Gillian, Karen Campbell, Diana Dalbotten, and Keisha Varma. "Cycles: A Culturally-Relevant Approach to Climate Change Education in Native Communities." *Journal of Curriculum & Instruction* 6, no. 1 (2012): 73–89.

Shea, Nicole A., Chrystalla Mouza, and Andrea Drewes. "Climate Change Professional Development: Design, Implementation, and Initial Outcomes on Teacher Learning, Practice, and Student Beliefs." *Journal of Science Teacher Education: The official journal of the Association for Science Teacher Education* 27, no. 3 (2016): 235–58.

Shea, Nicole, Chrystalla Mouza, and Andrea Drewes. "Climate Change Professional Development: Design, Implementation, and Initial Outcomes on Teacher Learning, Practice, and Student Beliefs." *Journal of Science Teacher Education* 27, no. 3 (2016): 235–58.

Sun, Yi, Roger C. K. Chan, and Huiwei Chen. "Learning with Geographical Sensitivity: Place-Based Education and Its Praxis." *The Professional Geographer* 68, no. 4 (2016): 574-83.

Supovitz, Jonathan A., and Herbert M. Turner. "The Effects of Professional Development on Science Teaching Practices and Classroom Culture." *Journal of Research in Science Teaching: The Official Journal of the National Association for Research in Science Teaching* 37, no. 9 (2000): 963–80.

Trott, Carlie D. "Reshaping Our World: Collaborating with Children for Community-Based Climate Change Action." *Action Research* 17, no. 1 (2019/03/01 2019): 42-62.

Williamson, Kathryn, Jamie Shinn, Deb Hemler, and Sandra M. Fallon. "A Case Study for Climate Change Teacher Professional Development in West Virginia." *Journal of Sustainability Education* 28 (2023).

Windschitl, Mark. "Inquiry Projects in Science Teacher Education: What Can Investigative Experiences Reveal About Teacher Thinking and Eventual Classroom Practice?". *Science Education* 87, no. 1 (2003): 112.

———. *Teaching Climate Change: Fostering Understanding, Resilience, and a Commitment to Justice*. Harvard Education Press, 2023.

Windschitl, Mark, and David Stroupe. *The Three-Story Challenge: Implications of the Next Generation Science Standards for Teacher Preparation.* Vol. 68, 2017. doi:10.1177/0022487117696278.

Windschitl, Mark, Jessica Thompson, and Melissa Braaten. *Ambitious Science Teaching.* Cambridge, MA: Harvard Education Press, 2018.

CHAPTER 21

Educational Policy & Climate Change Education in New Jersey

Sarah Sterling-Laldee

Environmental literacy has deep roots in New Jersey, nurtured at every step by committed public actors who have advocated tirelessly for new and improved public policy. In 1971, in response to increased public awareness of environmental challenges and the budding environmental movement spreading across the nation, the New Jersey legislature passed the Environmental Education Act, establishing district-based education programs and the Environmental Education Council. More than a decade later, in 1985, the newly formed Alliance for New Jersey Environmental Education began advocating to establish a statewide environmental education master plan to replace the original Environmental Education Act.

Several years later, in 1989, Republican governor Thomas H. Kean convened the New Jersey Commission on Environmental Education (NJCEE), which was charged with creating and implementing such a plan. The Commission, situated within the New Jersey Department of Environmental Protection (NJDEP), pulled together leaders from across a wide array of sectors of public life, from formal K–12 and higher education institutions to nonprofit and state agencies and religious leaders.[1] In 1993, the Commission released *Environmental Education in New Jersey: A Plan of Action*, which made policy recommendations meant to improve New Jersey residents' environmental literacy.

Up until 2019, the thought leadership and advocacy around environmental education rested squarely within the NJDEP and was largely supported by informal education advocacy groups, such as the Alliance for New Jersey Environmental Education and New Jersey Audubon. The NJCEE, supported by NJDEP staff, served as the major guiding influence on comprehensive environmental education efforts. Their reports and other publications offered recommendations for formal and informal education policy but lacked political teeth or funding to back up those recommendations. While environmental content was first introduced in the New Jersey Core Curriculum Content Standards in 1996, it was limited in scope and rested almost entirely within secondary science.

It took four more years before significant state dollars were allocated to support the implementation of the Commission's suggestions. This funding came by way of the NJDEP's FY 2000 Integrating the Environment as a Context for Learning grants, and Governor Christie Todd Whitman's Earth Day Education Fund. Even this funding was minimal, with roughly $200,000 to support efforts across the state. From that point until 2022, no significant state funds were allocated to support the implementation of the various recommendations put forth.

Environmental literacy was also largely viewed as a boutique issue, and in formal K–12 education, tangential to "real" learning challenges, particularly in urban settings. Until recently, environmentalism was not presented as an equity issue, and the agencies doing the work did not reflect the diversity of New Jersey communities or speak to their needs or challenges[2]. Communities like my own in Paterson, New Jersey, a former industrial hub and the third largest city in the state, face disparate impacts from climate change and other environmental mismanagement and neglect, but they weren't often represented by advocacy groups until after 2016.

Additionally, the accountability movement in formal education, which had begun in the 1980's but ramped up significantly with the No Child Left Behind Act (2008), further relegated environmental literacy (and science and civic literacy, as well) to the background when math and English language arts learning outcomes became explicitly tied to school funding. In New Jersey, these outcomes were also tied directly to teacher and administrator salaries through the Teach NJ Act in 2013. This move was again felt disproportionately in less affluent districts, both urban and rural, which faced much greater challenges in meeting the learning goals set out by accountability systems. The byproduct has been the curriculum's narrowing for many children, so they do not have the opportunity to engage in the socio-scientific learning foundational to their participation as informed citizens in a rapidly changing world.

With all its challenges, there have also been some major successes. Corporate philanthropy and private foundation funds have allowed for the expansion of partnerships and the development of model projects. This funding and the interagency partnerships that developed as a result supported initiatives across the state. I benefitted from this funding to support several pilot projects in Paterson, where I was a teacher and then supervisor of science instruction. External funding and cross-agency collaboration continue to be crucial for seeding innovation so that when state and federal funding do become available, we have rich models to scale up.

We've had much more momentum since 2019. In four short years, we've adopted standards that support climate change education across all nine content areas, allocated $5 million of state funding in fiscal years 2023 and 2024 to support K–12 classrooms, and created the Climate Change Education Unit in the New Jersey Department of Education to support that work.

So, what facilitated this change? I would argue social and political will. One contributing factor was the student climate movement, also known as Fridays for Future. Students across New Jersey participated in the Global Climate Strike and school walkouts in March 2019, as well as other climate actions throughout the rest of the year. This signaled to our leaders—including First Lady Tammy Murphy, a longtime supporter of climate action—that we could and should capitalize on the upcoming state learning standards revision cycle to thoughtfully embed climate change learning across all content areas up for revision because it mattered deeply *to students* across the state.

The first lady began to meet with the content revision teams to garner support for this initiative and in 2020, standards supporting climate-change education were adopted in fine and performing arts, comprehensive health and physical education, world languages, computer science and design thinking, career readiness, life literacies and key skills, social studies, and science. In the fall of 2023, our English language arts and math standards revisions were adopted with climate change companion guides.

We chose to focus on standards in New Jersey because our state constitution guarantees local control of schools. The state Department of Education does not mandate curriculum or engage in statewide textbook adoptions, as you might find in other states like California or Texas. We set the goals but do not dictate specific pedagogical approaches or learning resources to meet those goals, allowing individual districts to make decisions that best fit their populations. As you can probably imagine, though, it also means there is tremendous variation in how districts work to meet these goals.

Focusing our attention on learning standards also moves this work squarely into the formal education domain. Without specific guidance via standards, this learning was optional, which meant there were great disparities within and across districts in who had the opportunity to improve their climate literacy. As I mentioned before, students in the communities most adversely impacted by climate challenges were not offered opportunities to engage in learning experiences that would build the knowledge and skills they needed to address those issues now or in the future.

Moreover, the creation of the standards was meant to share the responsibility for climate literacy across content areas. Learning about the environment and climate has traditionally taken place almost exclusively in the science classroom. Science teachers (I was one for nearly fifteen years) had to hold all the water for this learning and attempt to coax their colleagues into interdisciplinary projects. We know from years of research on interdisciplinary, transdisciplinary, project-based, problem-based, and STEAM learning that students have outsized engagement for learning when it is connected across content areas and related to real-world issues in their local contexts. We also know students need early and regular engagement with STEAM content, especially if we are going to create successful, thoughtful contributors to society and our future economy, particularly in communities that may be underrepresented in these fields. So, creating and supporting the new standards' implementation is meant to push us toward these outcomes.

Like our peers across the country, we continue to be challenged by the politicization of climate science. We faced significant pushback on the adoption of the 2023 English language arts and math standards, so much so that we had to create companion guides rather than provide explicit examples in the standards themselves. We need to frame the conversation about these learning goals in a way that pulls them out of the political debate. Focusing on the future workforce and skills development is one way to do this, as long as we continue to situate the learning in ethics and civic responsibilities.

We also have challenges around ensuring *all* students have the foundational science literacy they need to participate meaningfully in society. Whatever our own schooling experiences may have been, our children need to arrive at secondary school more prepared to meet the challenges the world is already setting up for them. They need opportunities to engage with the natural world early and often, to develop systems thinking and data literacy skills. *All students need this*, irrespective of their first language, socioeconomic status, or perceived abilities. Too often, adults are acting as gatekeepers in ways that limit opportunities for children, even when they believe they have good intentions.

We also continue to struggle with a related perceived challenge: that there is not enough time in the school day, week, or year to engage in authentic, interdisciplinary learning because we must choose to spend our time on foundational gaps in English language arts and math skills. It is not an either/or proposition, but it does require us to rethink traditional scheduling; engage teachers in interdisciplinary, collaborative learning; and allow for things to get a little messy in the short term. I was so hopeful as we came out of the COVID-19 pandemic that we might reimagine the school days' structure and how we support teacher learning. Instead, most of us snapped back to pre-pandemic normalcy. This felt good in the moment but did not help us address the structural constraints that are of our own design. It also felt good to state and district leadership, not to teachers and students, which can be attested to with our ongoing struggles with chronic absenteeism among both groups and teacher attrition.

Our dynamic systems are evolving so fast that it is difficult for formal education to keep up. As new fields emerge, we will be challenged to identify, train, and credential teachers, particularly for career and technical education courses. We can't slow the rest of the world down, so we will have to become nimbler and rethink how we train and credential teachers. Lastly, as is so often the case, we have been working in silos. There is a need to pull together key stakeholders to align our work and build cohesion so our work is complementary rather than duplicative, or worse, competitive.

There is so much good work happening around the country, but we have limited opportunities to collaborate. Convenings like the Education Development Center's Preparing for a Green and Blue Workforce, the National Science Teaching Association's national conference focused on climate justice, and the great work Frank Niepold is doing at NOAA with the CLEAN Network are prime examples of ways we can begin to maximize reach. We need to share ideas and scale up successful, replicable work to quickly create a more climate-literate and resilient public.[3]

We are just beginning to explore emerging opportunities to create model programs for local districts and community colleges so they can partner to build New Jersey's talent and industry pipeline for the emerging green and blue economies. In New Jersey, these models are incredibly important for districts that may lack the capacity to build out their own programs. As a local-control state, we don't mandate programs of study, which can mean districts with more capacity can do more, while those that lack bandwidth miss opportunities to widen the playing field for their students. We are thinking deeply about how we can impact the trajectories available for students, particularly in communities that may not have had equitable access to high-quality STEAM learning and outdoor education or community-based role models in these fields. That starts as early as Pre-K, and is the reason standards, curriculum frameworks, and other state-level legislation is critical to students' future success in the workforce and their ability to contribute meaningfully to society.

Notes

1. New Jersey Environmental Education Commission (1994, 2000).
2. Smithsonian Institute (2017).
3. James Elder et al. *State Level Legislation Concerning K-12 Climate Change Education* (2023), NAAEE *Mapping the Landscape of K-1 Climate Change Education Policy in the United States* (2022).

Bibliography

Elder, James, Anisa Heming, and Jacqueline Maley, *State Level Legislation Concerning K-12 Climate Change Education* (Washington, DC: Center for Green Schools, June 2023). https://www.usgbc.org/resources/state-level-legislation-concerning-k-12-climate-change-education

New Jersey Environmental Education Commission, *Environmental Education in New Jersey: A Plan of Action*, 2nd ed. (Trenton, NJ: New Jersey Department of Environmental Protection, May 1994), doi:10.7282/T3416W9R

New Jersey Commission on Environmental Education. *Building Blocks for the Future: Environmental Education in New Jersey* (Trenton, NJ: New Jersey Department of Environmental Protection, 2000). doi:10.7282/T37S7N15.

North American Association for Environmental Education. *Mapping the Landscape of K-1 Climate Change Education Policy in the United States* (Washington, DC: NAAEE, May 2022). https://eepro.naaee.org/resource/mapping-landscape-k-12-climate-change-education-policy-united-states

Smithsonian Science Education Center, *Fostering Change: Ideas and Best Practices for Diversity in STEM Teaching in K-12 Classrooms* (Washington, DC: Smithsonian Institution, 2017). https://ssec.si.edu/fostering-change-ideas-and-best-practices-diversity-stem-teaching-k-12-classrooms

CHAPTER 22

Case Studies from Outside New Jersey

Christine Whitcraft, Kelley Lê, Kimi Waite, Jim Clifford, and Amara Ifeji

New Jersey has formalized its commitment to comprehensive climate change education through updated learning standards, but it is not the only state in the United States with energy surrounding climate education. This chapter provides vignettes from California, Connecticut, and Maine, three additional US leaders in climate-change education. Notably, each state took a different approach to expanding the reach of climate-change education. For example, in California, institutions of higher education have worked together to provide centralized climate education for pre- and in-service teachers. On the other hand, Connecticut took a legislative approach to include specific learning standards in science and social studies. Finally, Maine's action on climate education is centered on supporting teachers through professional learning experiences. Though each state approaches the integration of climate change into K–12 learning differently, the resulting shifts in instruction all address the same overarching shared goal: to enhance climate literacy through teaching and learning.

California

by Christine Whitcraft and Kelley Lê

With over six million students enrolled in its K-12 public schools, California is an ideal place to promote and scale new models in effective pedagogy for climate change. A recent report published by National Public Radio (NPR) revealed that more than 80 percent of US parents support teaching climate change, but 55 percent of teachers say they do not teach it because it is outside their content area.

[1] These findings, paired with anecdotal evidence from K–12 students and other environmental justice research influenced the formation of the Environmental and Climate Change Literacy Projects (ECCLPs). The mission of ECCLPs is to ensure all California public-school students are prepared to address the environmental challenges and inequities associated with worsening climate change. ECCLPs' core belief is that justice, equity, diversity, and inclusion efforts will be achieved by scaling up PreK–12 education initiatives around environmental and climate change literacy so California's youth become climate champions who are educated and fully capable of tackling climate change and environmental disruptions, now and moving forward. Formed in 2019 and re-launched in 2021, ECCLPs' aim is to refocus climate change and environmental literacy in schools and society by strengthening its presence both in the K–12 curriculum and in educator preparation programs across the state. ECCLPs is a collaborative effort wherein educators, higher education faculty, and researchers in the University of California (UC) and California State University (CSU) systems work with environmental advocates, policymakers, and nonprofits focused on integrating environmental and climate literacy across California's PreK–12 public school system. The UC system (ten campuses) and CSU system (twenty-three campuses) have partnered to establish a first-of-its-kind statewide center to advance climate change literacy for California youth and families. ECCLPs was envisioned to drive systems-level change that ensures California educators can prepare future generations of students to be effective environmental stewards and positions California as a leader in global efforts toward a sustainable planet.

Housing ECCLPs in the UC and CSU systems was an intentional choice, because combined, these institutions are the highest producers of California's PreK–12 teachers, conferring more than seven thousand teacher credentials every year and offering ongoing professional development to thousands of existing teachers. They play a crucial role in building the capacity of California's education system to teach about the environment and climate.

As ECCLPs works to achieve this mission, we center several core ideas. First, education, particularly within PreK–12 schools, can be a catalyst for climate and environmental actions, justice, and solutions. We center historically marginalized communities and peoples, as they contribute the least to carbon emissions fueling the crisis but unjustly suffer more from the impacts of climate change and environmental injustices exacerbated by the climate emergency. Because addressing climate change is a multifaceted challenge, developing culturally relevant solutions requires different subject-area expertise, as well as engagement with both systems and local community organizations at the forefront of the issues. Accomplishing these goals also requires that communities, including and prioritizing Indigenous cultures and marginalized communities, should be authentically included and actively engaged in addressing climate issues and developing solutions. In this way, our efforts seek to eliminate systemic barriers and inequities, while advancing equitable access to resources and opportunities. Additionally, building partnerships and sharing knowledge and resources can amplify the impact of our climate literacy efforts. With these values at the forefront, ECCLPs exists to advance climate and environmental justice literacy with and for PreK–12 schools to catalyze climate action and solutions.

WHAT ARE ECCLPS INITIATIVES?

Working with pre- and in-service PreK–12 teachers, our primary goal is to envision and broadly implement inter- and transdisciplinary approaches to climate literacy that:

- Focus on culturally relevant solutions;
- Activate community-driven action;
- Honor Indigenous ways of knowing and being; and
- Support those most impacted by the climate crisis.

Developing the measures needed to track our work's impact *will* take time. We continue to be transparent with our partners and supporters throughout that process. To improve our efforts to effectively and efficiently meet the needs of PreK–12 students and teachers, we aim to connect our initiatives with those of other leading organizations and institutions to amplify our efforts around climate literacy; share our learnings, research, and resources; and become a trusted local, statewide, and global partner.

An initial executive document (2019) laid out three specific initiatives as central to achieving ECCLPs' goals, and these continue to ground our efforts. However, as ECCLPs moved forward, committee members identified that these initiatives needed to be subdivided into smaller tasks with additional areas of focus. The first goal is for ECCLPs to engage with the California Commission on Teacher Credentialing and other state entities to integrate environmental and climate change literacy across relevant subjects and enhance teacher preparation and professional learning in environmental and climate literacy. To achieve this engagement, ECCLPs is gathering data on how teacher preparation currently approaches climate and environmental literacy, so we can understand existing gaps. ECCLPs will also work with specialists in education preparation programs to assist candidates who are focused on climate and environmental inquiry cycles in their credentialing process.

Next, ECCLPs aims to establish a bipartisan California State Taskforce for the broad promotion of environmental and climate change literacy for K–12 students, including partnerships with community-based organizations and youth groups. To enhance the efficacy of this taskforce, ECCLPs will identify a framework that centers on key practices and approaches to provide coherence for our state teacher education programs. Finally, ECCLPs will produce scientific and social science research to develop evidence-based approaches to advancing climate change literacy. In keeping with ECCLPs' core values, this research will recognize expertise and wisdom from multiple sources, including western knowledge (science, math, history-social science, etc.), Indigenous knowledge, and local communities. Committees focused on each of these goals have been established within ECCLPs to move the initiatives forward. UC Irvine serves as the main hub site for California's ECCLPs Center, with distributed satellite offices geographically dispersed across partnering campuses.

ECCLPS has already gained some traction on these initiatives. We co-hosted the California listening session with the National Oceanic and Atmospheric Administration (NOAA), which convened a variety of stakeholders to discuss and improve the next iteration of the national Climate Literacy Guide to be released in 2024.[2] Additionally, the ECCLPs UC Irvine campus task force hosted a kick-off event to foster connections and conversations between faculty/staff members across all departments and local Orange County teachers. Together with United States Global Change Research Program and NOAA, we brought together researchers across all fields, PreK–12 formal and informal educators, writers of the NCA5,[3] and other leaders at the American Geophysical Union conference to discuss the importance of the Fifth National Climate Assessment. The high school student survey results from this assessment are in, and we will share these reports widely in January 2025.

One California Educator's Journey

By Kimi Waite

When I worked as a kindergarten teacher in South Los Angeles, most of my five- and six-year-old students were quick to recognize the lack of green spaces where they lived. At the beginning of the school year, when the days were still hot from the summer sun, my students would often be exhausted after recess due to the heat, expressing, "There's no shade," "I want a tree to sit under," and "It's too hot." Several students also noted how "on TV, there's trees and cool slides on playgrounds." They asked if other neighborhoods and cities "have problems with no trees and parks" and why.[4] At a very young age, students are keenly aware of inequities around them, and my students' observations encouraged us to investigate issues of climate justice and environmental justice in our community, including access to green spaces and pollution.

Similar to the observations from my kindergarten class, environmental justice activists in California have long recognized the connection between climate policy and the everyday lived experiences of our state's most vulnerable residents. Activists have also long understood the urgency of climate action due to disproportionate impacts of climate change on frontline communities. One example is the San Joaquin Valley in California, the most productive agricultural region in the world, but "rising temperature and chronic air pollution make it an increasingly dangerous place to work,"[5] especially for farmworkers, as climate change has added increasing dangers to their essential jobs that feed the nation. California's Fourth Climate Change Assessment predicted that the valley is likely to see a sevenfold increase in extreme heat days.[6]

Teaching about environmental hazards without addressing the related social, economic, and political forces that dictate policy and action will make environmental education programs virtually irrelevant for students from frontline communities, who are disproportionately students of color. Educating for environmental justice demands an interdisciplinary viewpoint and approach, positioning educators across all grades and subjects to contribute meaningfully.[7] This makes the work of the UC-CSU Environmental and Climate Change Literacy Projects program incredibly important to advance PreK–12 climate and environmental literacy, justice, and action.

Connecticut

By Jim Clifford

Connecticut began its climate education journey by developing an Environmental Literacy Plan in 2010.[8] This was a good first step, but it had no immediate impact on Connecticut schools. Similarly, Connecticut's adoption of the Next Generation Science Standards (NGSS) in 2015,[9] codified in Section 10-16b of the Connecticut General Statutes, provided the *opportunity* to teach about climate change in Connecticut schools, but did not *require* it: "Sec. 10-16b. Prescribed courses of study. In the public schools the program of instruction offered shall include at least the following subject matter, . . . science, which *may* include the climate change curriculum consistent with the Next Generation Science Standards."

In 2019, Christine Palm, a freshman legislator in the Connecticut General Assembly, discovered climate change was not being taught universally in Connecticut public schools, and she committed to change that. The first hurdle was convincing the General Assembly's Education Committee chairs that climate change was *not* already being taught. Strong pushback from the state's Association of School Superintendents had convinced the committee chairs that all schools were already engaged in such teaching and that the bill was therefore not needed. However, Palm had heard from many young people that their science curriculum either omitted climate change altogether or taught it inadequately.

A comprehensive survey of every school in Connecticut's 169 towns was not feasible, because state legislators only serve part-time and have limited staff. Instead, Palm turned to a young climate activist named Sena Wazer, who had previously approached her at the State Capitol, saying, "I'm a 15-year-old climate activist and I want to help." Shortly thereafter, Wazer and others in the Connecticut chapter of the youth-led Sunrise Movement organized a climate justice rally at the State Capitol in Hartford. Several thousand young people turned out. Palm and Wazer prepared a simple survey that asked three questions:

1. Does your school's science program teach climate change?
2. If so, was it adequate, in your opinion?
3. Do you believe climate change (as primarily a human-made problem) should be taught in all public schools?

The overwhelming response was that climate change education was spotty at best and sorely needed. A pattern soon appeared: Underserved communities—in Connecticut, those consist primarily of black and brown people—were the least likely to learn about climate change in school, despite the well-known fact that these communities are more burdened by environmental issues such as pollution and polluting facilities, loss of tree canopy, fewer recreational spaces, and less access to clean waterways. This is not unlike the trends that emerged in California, as noted earlier in this chapter.

The next step was marshaling this youthful energy into an army of young citizens willing to testify on Palm's bill requiring climate change education at a public hearing held by the Education Committee. With the help of Rep. Geraldo Reyes (D-Waterbury), activist and educator Leticia Colon de Mejias, and several high-school science teachers, Palm was able to enlist about seventy-five testifiers, more than five times the number of people who typically turn out for a public hearing. Even hardened lawmakers were moved to see so many young people waiting patiently in line to express their despair over climate change, their anger over governmental inaction, and their hope that such a simple bill could have a lasting impact.

Eventually House Bill 5285[10] passed, requiring climate change curriculum in all science classes by changing one word in C.G.S. Section 10-16b: To wit, "science, which *may* include the climate change curriculum consistent with the Next Generation Science Standards" to "science, which *shall* include the climate change curriculum." An eight-hour filibuster by the Republican caucus, whose members argued that "carbon is good for plants," that their constituents have the "right to drive gas-guzzling cars in America," and that cold winter days "prove climate change is a hoax," failed to derail the bill. Thus, Connecticut became the second state in the nation (after New Jersey) to require climate change instruction in science classes.

People who do not work in the public policy field can hardly comprehend that changing one word in a statute could take four years and seven legislative sessions to pass both the House and Senate and be signed into law by the governor. Palm, whose efforts were acknowledged when she received the Walter Cronkite Award for Climate Education, urged her young allies to continue their leadership in the climate movement: "To all the young people just starting out in this movement, please don't give up on us. Please run for office. You are leading the way now and you're going to have to hold your lantern up just a little bit higher because too many of us are still in the dark, struggling to see what you have known all your young lives."[11]

In addition to these changes in science education requirements, Connecticut also expanded its social studies coverage of climate change and environmental literacy. The adoption of the College, Career, and Civic Life (C3) Framework for Social Studies State Standards in 2015 opened the door to climate change instruction with one of its geography themes: the "human-environment interaction." The C3 Framework, with its emphasis on an "inquiry arc," shared some of the processes set forth in the NGSS, but it lacked specific content standards for climate change or environmental literacy.

Connecticut's first set of social studies standards was initiated by Public Act 21-2ss, which called for a K–8 model curriculum. The legislation required the model curriculum to include and integrate a number of topics prescribed in Section 10-16b of the General Statutes, including climate change. The design and development of exemplary model curricula required new social studies standards. Connecticut teachers, curriculum leaders, and administrators supported the need for new standards, although climate change was not a top priority.

One teacher, Jim Clifford of Amity Regional High School, vigorously advocated for the inclusion of climate change as a member of the state's Steering Committee and as a standards writer. For a year and a half, Clifford researched, wrote, and revised climate change, sustainability, and environmental literacy standards for K–12 economics, civics, geography, and history. In October 2023, the Connecticut State Board of Education unanimously adopted the new social studies content standards, which includes some of the most detailed climate change and environmental language in the nation. An excerpt of these standards is attached as Appendix X.

Of note, the new state social studies climate-change standards, though instructive and intended to shape social studies curriculum, are not mandatory, as is Connecticut's science curriculum. The standards clearly state, "In Connecticut, each district has local autonomy to use the standards to develop a social studies curriculum that is relevant and accessible to their school, local, or regional community."[12] Nevertheless, the revised standards signal a new statewide emphasis to include climate education in subjects besides the sciences.

One challenge facing Connecticut science and social studies teachers is the lack of funding for climate change professional development training and educational materials. Fortunately, this void has been filled in part by nonprofits like SubjectToClimate,[13] which is building a statewide climate education hub in Connecticut, like the ones it has created in New Jersey, Maine, Wisconsin, and Oregon.

TAKE-AWAYS READERS CAN USE IN THEIR OWN STATES AND CONTEXTS

Although most states lack the political support to pass legislation on climate-change education, educators can pursue other avenues to teach about climate change. A majority of states have adopted environmental literacy plans that provide a framework and opportunity for school districts and teachers to create, expand, and improve environmental education programs that incorporate climate change education. Most states have adopted the NGSS and the C3 Framework or incorporated changes in their science and social studies curricula based on these standards in school districts across the nation.[14] The focus on inquiry in both disciplines paved the way in Connecticut for studying climate change in science and social studies, and it may also empower intrepid teachers in other states.

Individual state legislators, teachers, and particularly students, can play key roles in demanding climate change curriculum in their states. One approach is to argue for standards and content related to the terms *environmental literacy*, *conservation*, and *sustainability*, which are less polarizing and, in some cases, more appropriate than *climate change*. For states that are not able or willing to fund climate-related curriculum, resources, and professional training, nonprofits like SubjectToClimate, NASA, and National Geographic provide excellent resources that can be leveraged in any classroom context.

Maine

By Amara Ifeji

"What do we want? Climate justice! When do we want it? Now!" my friends and I enthusiastically exclaimed as we stood among a crowd of fellow students gathered at our community park in Bangor, Maine. Like millions of other youths worldwide, we answered the call to action spearheaded by the Fridays for the Future[15] climate movement to advocate for the planet. In doing so, we also joined a statewide effort led by Maine students across universities, schools, and communities, which spearheaded youth-led coalitions that built Maine's burgeoning youth climate movement.

As a budding environmental educator, I led stormwater monitoring efforts with fellow peers at my high school. This work enabled me to recognize the importance of education in environmental advocacy. However, I wondered how I would scale my education initiatives to positively impact students beyond my community. Little did I know I would soon connect with folks across the state who similarly regarded education as a critical tool for social change, nor did I realize our advocacy efforts would ultimately culminate in the Maine Legislature investing more than $2 million in a historic K–12 climate-education program.

I joined the Nature-Based Education Consortium and its Climate Education Advocacy working group when the pandemic shifted Maine's youth climate movement from in-person rallies to Zoom meetings. Especially during the isolating early months of the pandemic, I was grateful to meet with people across Maine who shared my vision to enhance climate and environmental awareness for Maine students. Our intergenerational, youth-led group of students, educators, and community leaders bonded around a shared background: lacking climate education in our public school experience.

Although we were eager to jumpstart our advocacy, we recognized that investing time in building our new community would ultimately enable us to achieve our goals. We spent the first few months of our time co-creating group working agreements and articulating our unique reasons for contributing to this work. Together, we fostered a culture of mutual respect and a way of being that values all contributions to the collective work.

The end of 2020 presented an opportunity for our group to implement our shared visions. We met with the governor's office, which coordinated the newly formed Maine Climate Council. The council delivers a climate action plan to the governor and legislature every four years. Entitled "Maine Won't Wait,"[16] the plan comprehensively outlines actionable strategies to mitigate climate impacts. However, we recognized climate education was an all-but-forgotten component of the inaugural plan. Together, we wrote recommendations to include climate education as part of "Maine Won't Wait" and were thrilled to see these recommendations in the completed plan.

Upon our success, we realized we had the momentum for a statewide climate education initiative. The 2019 Census for Community-Based Environmental Learning[17] cited climate education as the number one community-based environmental learning topic for which Maine educators requested additional resource support. Given this data, we set out to engage with stakeholders to better understand Maine's current landscape of climate education. As part of my role with the Maine Environmental Education Association, I served as the lead coordinator for the state's first Climate Education Summit in June 2021.

Over two hundred youth and adult leaders from across various sectors and all parts of the state came together around a shared vision to advance equitable access to climate change education in Maine. I supported a diverse co-planning team of more than twenty individuals led by adult and youth co-leads who organized the summit, which was held virtually over two weeks. Participants engaged in "vision labs" designed to identify critical resources, common barriers, and solutions to advance climate educator support, policy, and workforce development. Folks in the working group and beyond were left with actionable steps to realize a shared vision for climate education in Maine.

That summer, the Climate Education Advocacy working group had a retreat to strategize our next steps post-summit. As we discussed our ideas over tacos, we realized an opportunity to present climate change education as a strategy to address the goals outlined in the climate action plan. At the time, Maine was experiencing an unprecedented budget surplus and potential change in administration, so it was now or never. At the end of our time together that day, we committed to partnering with climate advocate Representative Lydia Blume to put forth legislation in the 2022 Maine Legislature. Modeled after Washington State's ClimeTime program,[18] we sought to establish a grant program to provide educators with funds to seek climate education professional development they could bring back to their classrooms.

Our first hurdle was the Legislative Council, which reviewed all the bills submitted to be heard in that session. Over seventy percent of bills were denied by the council, including ours. Although we thought that was the end of our advocacy efforts, Rep. Blume encouraged us to try to overturn the decision. So, we organized in less than a week to coordinate a teach-in for the legislative council. Over eighty individuals participated in the half-hour lunch-and-learn, which clarified the bill's aims and resulted in the council overturning their previous vote.

A deep sense of trust within our coalition was paramount as we approached the next hurdle: the bill's public hearing. Many in our group, including myself, had never attended a public hearing or testified for a bill. However, we were supported by adult advocates and coached on how to speak to legislators. Over ninety pieces of positive testimony resulted in the bill making it through the education committee with a favorable vote.

The bill garnered a broad network of outdoor engagement and environmental education organizations, and it was selected as a priority legislation for the Maine Environmental Priorities Coalition. Similarly, organizations in the Maine Youth for Climate Justice coalition organized to mobilize over one hundred students, educators, and legislators to the Statehouse in Augusta to support the bill. These efforts contributed to the bill's successful passing in the Maine House

CASE STUDIES FROM OUTSIDE NEW JERSEY 295

and Senate. As such, our working group and broader coalition of community partners rallied to tackle the last hurdle: the Appropriations Committee.

Meetings with members of the committee highlighted the importance of intergenerational organizing. The youth in our working group knew how to share the bill's story with legislators, ensuring that we highlighted our research-driven, community-informed process. However, we needed connections to set up meetings with the committee members. As such, adult advocates in the group organized the meetings for us and were present, but they gave us the space to share, only chiming in when we asked. Again, building trust was crucial.

Legislative Document 1902[19] was signed into law by Governor Mills on May 9, 2022, allocating $2.1 million to advance climate education in Maine public schools. To date, a climate education coordinator has been hired at the Maine Department of Education, and the department has granted funds to an inaugural cohort of schools, with two additional funding opportunities on the horizon. Moreover, many climate education curricular resources have been generated to improve climate literacy in Maine, including Maine's Climate Education Hub.[20] Led nationally by SubjectToClimate and locally by the Maine Environmental Education Association, the project convened Maine educators to compile standards-aligned, place-based educational resources for teachers, all of which are available online at no cost.

Above all else, Maine's climate education story is one of the successes that follow when centering those most impacted by decisions in generating solutions.

Figure 22.1   Governor Mills signing Legislative Document 1902

Our coalition partners and working group primarily comprised teachers who would teach climate-change topics and youth like me who felt underserved by climate change education, even though we are experiencing and will inherit the issue. As such, these perspectives were centered in our stakeholder engagement, legislation drafting, and advocacy initiatives.

Maine's story also illuminates the importance of community power. Our initiative did not just live with one organization, but it also built social infrastructure through a grassroots movement and diverse coalition of students, teachers, legislators, community leaders, and others. These individual voices uniquely contributed to the movement's success because every person carried out a role that suited their talents. Only through this community could we appropriately leverage individuals' strengths to amplify our impact.

Lastly, our initiative highlights the importance of research-backed storytelling. When communicating the program's importance to legislators, we made sure to underscore the participatory, community-driven Climate Education Summit, where much of the bill's language had originated. We also noted the climate education resource needs cited in the 2019 Census for Community-Based Environmental Learning. Furthermore, we told stories about the bill that would uniquely resonate with different legislators. For some, we highlighted the bill's workforce connections—students are more likely to stay and work in Maine if they know about our climate workforce offerings. For others, we stressed the importance of supporting our teachers. These research-backed, tailored narratives were crucial in having legislators recognize the bill's importance.

LD1902 and numerous initiatives have solidified Maine as a national leader in climate education. I hope sharing our story will inspire others to answer the call to action to advocate for the planet while leveraging education as a critical tool for social transformation.

Notes

1. Anya Kamenetz, "Most Teachers Don't Teach Climate Change; 4 in 5 Parents Wish They Did," NPR, April 22, 2019, https://www.npr.org/2019/04/22/714262267/most-teachers-dont-teach-climate-change-4-in-5-parents-wish-they-did.

2. Climate Literacy: The Essential Principles of Climate Sciences, 2009, United States Global Change Research Program. https://www.climate.gov/teaching/climate

3. NCA5 is a comprehensive teaching tool while emphasizing the importance of education to catalyze the recommended actions.

4. Kimi Waite, "Action Research for Environmental Justice In the Kindergarten Classroom," *Rethinking Schools*, 2022, https://rethinkingschools.org/articles/action-research-for-environmental-justice-in-the-kindergarten-classroom/

5. Liza Gross and Peter Aldhous, "Excessive Heat and Air Pollution Are Putting Farmworkers' Lives at Risk," *Mother Jones*, December 31, 2023, https://www.motherjones.com/politics/2023/12/heat-and-air-pollution-putting-scores-of-california-farmworkers-lives-at-risk/

6. James H. Thorne, Joseph Wraithwall, and Guido Franco, *California's Fourth Climate Change Assessment* (Sacramento, CA: California Natural Resources Agency, 2018), https://www.climateassessment.ca.gov/state/overview/

7. Kimi Waite, Teaching for environmental justice: Learning from the film "Manzanar, Diverted: When Water Becomes Dust." *Social Education* 8(3) 2024 [In Press].

8. Connecticut Outdoor and Environmental Education Association, *Connecticut's Environmental Literacy Plan, 2020-2025* (Woodbury, CT: COEEA, 2020). https://www.coeea.org/_files/ugd/3b1515_a503742496254b7c9b18eacfbcd9de9a.pdf?index=true

9. For more on the NGSS's attention to climate change, see Chapter 1 of this book.

10. An Act Concerning the Public School Curriculum, Bill No. 5285, State of Connecticut General Assembly, Committee on Education (2022), https://www.cga.ct.gov/2022/TOB/H/PDF/2022HB-05285-R00-HB.PDF

11. The Stone Soup Leadership Initiative, "Congratulations Award Recipients: Cronkite Award for Climate Education," https://myemail.constantcontact.com/Congratulations-to-the-Institute-s-Cronkite-Climate-Education-Award-Recipients.html?soid=1106306587764&aid=pi8_Dd6VUbI

12. Connecticut State Board of Education, *Connecticut Elementary and Secondary Social Studies Standards* (Hartford, CT: CSBoE, October 4, 2023), https://ctsocialstudies.org/images/downloads/Connecticut_Elementary_and_Secondary_Social_Studies_Frameworks/ct_social_studies_standards_approved10.4.2023.pdf

13. www.subjecttoclimate.org

14. S. G. Grant, John Lee, and Kathy Swan, "The State of the C3 Framework: An Inquiry Revolution in the Making," *Social Education* 87, no. 6 (2023), https://www.socialstudies.org/system/files/2023-12/se-870623361.pdf; Ray Bendici, "How States and School Districts Are Adopting the Next Generation Science Standards," *District Administration*, October 11, 2019, https://districtadministration.com/how-states-and-school-districts-are-adopting-the-next-generation-science-standards/

15. https://www.Fridaysforfuture.org

16. Maine Climate Council, *Maine Won't Wait: A Four-Year Plan for Climate Action*, (Augusta, ME: Maine Climate Council, December 2020), https://www.maine.gov/future/sites/maine.gov.future/files/inline-files/MaineWontWait_December2020.pdf.

17. Maine Math and Science Alliance, Maine Environmental Education Association, and Elmina B. Sewall Foundation, *Census of Community-Based Environmental Learning in Maine*, ed. Kate Kastelein (Augusta, ME: Maine Math and Science Alliance, 2019).

18. https://ospi.k12.wa.us/student-success/resources-subject-area/science/climetime

19. "Resolve, to Establish a Pilot Program to Encourage Climate Education in Maine Public Schools," HP 1409, 130th Maine Legislature, Second Regular Session (2022), https://legislature.maine.gov/bills/getPDF.asp?paper=HP1409&item=1&snum=130.

20. https://maineclimatehub.org/

CHAPTER 23

Climate Change in the Garden State's Science Standards

Glenn Branch, National Center for Science Education

State science standards play a central role in science education in US public schools. By establishing learning goals for students over the course of their K–12 science education, they dictate textbook content, provide the basis for statewide testing, influence pre-service teachers' coursework and in-service teachers' professional development, and supply the structure for science curricula constructed or chosen by local districts, thus setting the agenda for the day-to-day lesson plans devised by individual science teachers.

Moreover, especially regarding a socially controversial topic such as climate change, state science standards can provide a shield for teachers facing complaints about curriculum and instruction. Such complaints are not uncommon: In a 2014–15 national survey, about six percent of public middle and high school science teachers who taught about climate change reported experiencing pressure not to do so.[1] Teachers are more likely to be able to deflect such complaints when they can explain that the state expects them to teach in accordance with the standards.

The way state standards address climate change plays a significant role in determining what is presented about climate change in the public school classroom. Accordingly, *Making the Grade? How State Public School Standards Address Climate Change*, a 2020 report from the National Center for Science Education and the Texas Freedom Network Education Fund, systematically examined the treatment of climate change in middle and high school state science standards across the country.[2] Unsurprisingly, there was both good and bad news in the report.

The good news is that many states earned a B+ or better for how their standards address climate change overall. Massachusetts and the twenty states (and the District of Columbia) that adopted the Next Generation Science Standards[3] received a B+. Four states—Alaska, Colorado, New York, and North Dakota—

received an A-, while a single state, Wyoming, received an A. It is mildly surprising that Alaska, North Dakota, and Wyoming, with their strong economic dependence on the fossil-fuel industry, fared so well.

The bad news is that of the remaining twenty-four states, twenty received a C+ or worse, and ten received a D or worse—including some of the most populous states in the country, like Florida, Pennsylvania, Ohio, and Texas. Six states—Alabama, Georgia, Pennsylvania, South Carolina, Virginia, and Texas—received a failing grade. Significantly, the six states where the standards are not based on the framework upon which the NGSS are based[4] tended to fare badly: North Carolina received a C-, while the remaining five received grades of D or F.

Not addressed in the report, but worth considering, is the quality of the treatment of climate change in state science standards, considered nationally. A national grade can be calculated by weighing the individual state grades based on student enrollment in each state. In other words, if a US public school student were selected at random, what would the expected grade of his or her state's science standards be? The answer is C+ (a grade shared by Idaho, Nebraska, and Utah), reflecting the impact of the very low grades received by populous states such as Florida, Pennsylvania, and Texas.

New Jersey earned a B+ in the report, reflecting its adoption of the NGSS in 2014. As the research for the report was under way, however, New Jersey adopted a revised set of science standards as part of the effort initiated by Tammy Murphy to incorporate climate change throughout all its state education standards.[5] The new science standards are still recognizably the NGSS, but several paragraphs of guidance about teaching climate change were added. In light of the well-documented lack of preparation for science teachers to teach climate change, these paragraphs are welcome.[6]

Furthermore, the revised standards insert climate change into several earth and space science performance expectations from kindergarten to high school. For example, they expect fourth-graders to "[g]enerate and compare multiple solutions to reduce the impacts of natural Earth processes and climate change have [sic] on humans," while the corresponding passage of the NGSS mentions only "natural Earth processes."[7] The added emphasis on climate change encourages science teachers to present climate change whenever possible.

As a result, if the *Making the Grade?* study were to be conducted again, New Jersey would probably receive an A- or even an A, reflecting the added emphasis on climate change in its current science standards. New Jersey is not the only state to have improved its standards since the study was conducted: Pennsylvania, which previously earned a failing grade, would now probably receive a C; South Carolina, which previously also earned an F, would now probably receive a B+; and Indiana, which previously earned a D, would now probably receive an A-.

Since the study considered only science standards, the incorporation of climate change throughout New Jersey's state education standards, from career readiness to world languages, would not affect the state's grade. Because these new standards were implemented only in the 2022–23 school year, there are not yet data on their impact, but it is plausible to suppose that the presence of climate change across standards for all subjects helps increase student engagement and affords the opportunity for interdisciplinary collaboration between teachers, both of which improve the quality of climate-change education.

Also noteworthy is New Jersey's commitment to preparing its teachers, whether in science or not, to comply with the new climate-change-focused standards. The state allocated $5 million to support climate change education grants to the state's public schools in 2023 and a further $5 million to support climate change education in 2024.[8] New Jersey thus succeeded, as a supporter of these efforts quipped in a discussion of the need to prepare its educators to teach climate change, in "meeting the greenhouse effect with the greenback dollar."[9]

A salutary further step would be to increase the availability of high school classes in earth sciences and environmental sciences throughout New Jersey, and perhaps to require such a course for graduation. At present, such classes are usually not required for graduation from high school. As a result, according to the most recent data, only about twenty-three percent of US public high school graduating students take these courses, while practically all take a biology class.[10] But it is precisely earth science and environmental sciences classes that afford the best opportunity in K–12 education to present a thorough treatment of climate change.

A large majority, about 75 percent, of Americans agree schools should teach our children about the causes, consequences, and potential solutions to global warming.[11] State science standards are a primary determinant of whether public schools will succeed in doing so. New Jersey deserves credit for its efforts not only to improve its standards but also to support its public school teachers to ensure they will be able to comply with the new standards' demands. There is a long way to go, however, before the American educational system is up to the task of preparing today's students to cope with tomorrow's warmer world.

Acknowledgments

Portions of this chapter were adapted from Glenn Branch and Lin Andrews, "Climate Change in State Science Standards: A 2020 Snapshot," *In the Trenches* 11, no. 2 (2021): 1–5. Blake Touchet aided in assessing the current New Jersey state science standards, and Julia T. Simms and Lauren Madden offered helpful suggestions.

Notes

1. Eric Plutzer et al., "Climate Confusion among U.S. Teachers," *Science* 351, no. 6274 (2016): 664–65.
2. National Center for Science Education and Texas Freedom Network Education Fund, *Making the Grade? How State Public School Standards Address Climate Change*, 2020, https://climategrades.org/.
3. NGSS Lead States, *Next Generation Science Standards: For States, By States* (Washington, DC: National Academies Press, 2013).
4. National Research Council, *A Framework for K–12 Science Education: Practices, Crosscutting Concepts, and Core Ideas* (Washington DC: National Academies Press, 2012).
5. John Mooney, "State Board of Ed OKs New Teaching Standards for Climate Change, Sex Education," *NJ Spotlight News*, June 5, 2020, https://www.njspotlightnews.org/2020/06/state-board-of-ed-oks-new-teaching-standards-for-climate-change-and-sex-education/.
6. Plutzer et al., "Climate Confusion."
7. Compare 4-ESS-3-2 of New Jersey Department of Education, *2020 New Jersey Student Learning Standards, Science, Kindergarten through Grade 12* (https://www.nj.gov/education/standards/science/Docs/NJSLS-Science_K-12.pdf) to 4-ESS-3-2 of NGSS Lead States, *Next Generation Science Standards: For States, By States* (Washington DC: The National Academies Press, 2013).
8. Seyma Bayram, "New Jersey Requires Climate Change Education. A Year in, Here's How It's Going," National Public Radio, August 21, 2023, https://www.npr.org/2023/08/20/1191114786/new-jersey-requires-climate-change-education-a-year-in-heres-how-its-going. California, Maine, and Washington have also passed legislation funding professional development on climate change for science teachers.
9. Glenn Branch, "Follow-up Needed on NJ's Renewed Commitment to Climate Change Education," *NJ Spotlight News*, June 24, 2020, https://www.njspotlightnews.org/2020/06/op-ed-follow-up-needed-on-njs-renewed-commitment-to-climate-change-education/.
10. J. Brown et al., *Paths through Mathematics and Science: Patterns and Relationships in High School Coursetaking* (Washington, DC: National Center for Education Statistics, 2018), 16. https://nces.ed.gov/pubs2018/2018118.pdf.
11. Anthony Leiserowitz et al., *Climate Change in the American Mind: Politics & Policy, Fall 2023* (New Haven, CT; Fairfax, VA: Yale University and George Mason University, 2023), 42. https://climatecommunication.yale.edu/wp-content/uploads/2023/11/climate-change-american-mind-politics-policy-fall-2023.pdf.

Index

AASL. *See* American Association of School Librarians
Abidi, Maliha, 143
absenteeism, 282
abstract concepts, EfCA and, 66
academic achievement, student engagement relation to, 77
academic libraries, 160–61
Academic Search Premier, 156
Acadia National Park, 253
accountability, 62, 280
Acomfrah, John, 137
action-oriented learning, 215, 216, 221. *See also* problem-based learning
active listening, 244–45
activism. *See* climate activism
ACUA. *See* Atlantic County Municipal Utilities Authority
adaptation, 125, 127; ethical considerations for, 128; for PBL, 174
Advanced Placement courses (AP), 173
advocacy, 125, 244, 293; art and, 132; communication for, 69; student, 215
agricultural engineering, 44
agriculture industry, of New Jersey, 8, 13
air quality, 9; living wall relation to, 231; in San Joaquin Valley, 289
ALA. *See* American Library Association

Allen, Lauren B., 54
Alliance for New Jersey Environmental Education, 279
American Association of School Librarians (AASL), 155
American Geophysical Union, 288
American Library Association (ALA), 154
American Museum of Natural History, 217
American Psychological Association, 180
American Public Health Association (APHA), 239
American Sign Language (ASL), 33
anchor phenomenon, 202
Anderson, Tom, 137
"animal crossing," 89, *91*
animal habitats, 89
Antarctica, 8
anthropogenic forces, 28, 31, 39, 243
anxiety. *See* climate anxiety
AP. *See* Advanced Placement courses
APHA. *See* American Public Health Association
aquaculture industry, 12
aquatic ecosystems: invasive species in, 14; storms relation to, 11–12
aquifers, 10
Arctic, 7
argumentation, evidence-based, 129

art, 148; emotions in, 146; empathy and, 132, 147; endangered species, 142–43, 145; for environmental justice, 133, 134, 140, 141; public educational, 242; transdisciplinary approaches and, 149; upcycling materials in, 139
art club, 144
Art & Ecology Colloquium, at Ohio State University, 135
Art Education (journal), 135
Art Education for a Sustainable Planet (Bertling), 137
Artichoke Dance Company, 141
Arts Ed NJ, 241
Arts Education and Social and Emotional Learning Framework, 241
ASL. *See* American Sign Language
assisted migration, 15
Association of School Superintendents, 289–90
asthma, 181
Atlantic Coastal Plain, 13
Atlantic County Municipal Utilities Authority (ACUA), 241
atmosphere, 34, 37; biodiversity relation to, 140
Atmosphere Investigation, GLOBE, *105*
audits: by Rutgers Cooperative Extension, 229; for sustainability, 175
aviation, 87–88

Banksy, 136
Basche, Andrea, 57–58, 63
Battle, Colette Pichon, 218
beaches, 168, 250
Bear Tavern Elementary (BT), 224, 233
bee population, 218, 221n10
Bell, Philip, 237
Bentz, Julia, 137
Bertling, Joy, 137, 138
bias, resources and, 212–13, 220
Biblioteca EPM, 153
bicycle model metaphor, *55*, 56, 62, 70; critical thinking in, 54–55; emotions in, 67–68; in middle schools, 61; practical action in, 66
Big History, 216

"big picture," 87
Bill of Rights, 122, 125
biodiversity, 13–15, 37, 121, 123, 252, 257; atmosphere relation to, 140; Indigenous people relation to, 142; outdoor learning and, 45; in pollinator gardens, 226; STEM relation to, 80–81
bird migration, 239
Black youth, 181
Blandy, Douglas, 135
Bloom, Mark, 263
Blume, Lydia, 294
board of education, 96
Board of Education, of New Jersey, 3, 29
Boy and Girl Scout Projects, 234, 235
Boyer, Brenda, 154
Boy Scouts of America. *See* Scouting America
Bradley Beach, 168
Breslyn, Wayne, 266
BT. *See* Bear Tavern Elementary
Buhr, Susan M., 266
Bullard, Robert, 134
Byron, Pete, 242

C3. *See* College, Career, and Civic Life; College, Career, and Civic Life Framework for Social Studies State Standards
cafeterias, 69; compost in, 65, 85, 229–30; makerspaces in, 171
California, 285–86; environmental justice in, 288–89
California Commission on Teacher Credentialing, 287
California State Taskforce, 287–88
California State University (CSU), 286
Cantell, Hannele, 54, 67
Cape May Vocational School, 242
carbon footprints, 78, 94; of families, 252
career readiness, 213, 220
Career Readiness, Life Literacies and Key Skills (standards), 83–84
carousel walk, 169
CARS rubric, 127; SSI and, 128
Census for Community-Based Environmental Learning, 293, 296

INDEX 305

Center for Mathematics, Science, and Computer Education (CMSCE), 166; professional development and, 175
Chesapeake Bay Watch, 215
Chickadee Creek Farm, 224
Childish Gambino, 141
Children and Nature Network, 46
Chin, Mel, 134
citizen science projects, 215
civic action, social studies and, 120
civic engagement, 57, 83; in cafeterias, 85; in PBL, 168
Civic Ideas and Practices, in NCSS, 118
Civil Conservation Corps, 14
claim-based statements, 77
class discussions, 21
Classroom Earth, 38
classroom organization, 173
CLEAN Network, 283
clean-up day, 45
Clifford, Jim, 291
climate. *See specific topics*
Climate Academies, 270
climate activism, 50, 257; climate anxiety relation to, 254–55; field trips and, 87; social media for, 255–56; social studies and, 78
climate anxiety, 30, 179, 180, 253–55; coping strategies for, 181, 185; empathy for, 187; SEL and, 87
Climate Awakening, 183
The Climate Challenge, 239
climate change. *See specific topics*
"climate change book of the week," 48
Climate Change Chicken Coop Project Rubric, *90*
Climate Change Collage, *132*
"climate change corner," 91, 92
Climate Change Council, 158–59
climate change deniers, 159
Climate Change Education Unit, 280–81
Climate Change Project, 58–59, 63–64; collaboration in, 68
climate doomism, 254–55
Climate Doom to Messy Hope Handbook, 188
Climate Education Advocacy, 293, 294
Climate Education Hub, of Maine, 295

Climate Education Summit, 293, 296
Climate Generation, 242
climate hero portrait, *135*, 142, 145
Climate Initiative, 255
Climate Literacy Guide, 288
climate policies, 53, 122, 251, 256
Climate Reality Project, 3
climate resilience, 39, 43, 125
Climate Writer in Residence, at West Vancouver Memorial Library, 154
climate zones, 37
ClimeTime program, 294
clubs, 237
CMSCE. *See* Center for Mathematics, Science, and Computer Education
coastline, 7, 12
CODAP. *See* Common Online Data Analysis Platform
collaboration, 47, 106; in Climate Change Project, 68; with community organizations, 183, 223–24; in design thinking, 90, 167; digital tools for, 214; with experts, 58; in gardens, 226–27; with Green Team, 96; in inquiry, 176; interdisciplinary, 56, 59, 66–67, 70, 120; for large-scale data collection, 49–50; on living wall, 231; in makerspaces, 159–60; with nonformal education, 240; in problem-solving, 63, 167, 264; in science, 116; with teachers, 89, 171, 301
Collaborative Data Collection Display, *49*
collaborative message board, 47
collective action, 67, 68; EfS and, 194; mental health relation to, 189
collective decision-making, 54
collective efficacy, knowledge and, 54
collective vision, 267
College, Career and Civic Life Framework for Social Studies State Standards (C3), 118, 291, 292
The College of New Jersey (TCNJ): PDSN of, 19; Sustainability Institute at, 2
college readiness, 213, 220
Colon de Mejias, Leticia, 290
Common Core State Standards: for mathematics, 56–57; NGSS relation to, 2; NJSLS relation to, 117; PL and, 266

306 INDEX

Common Online Data Analysis Platform (CODAP), 106
communication, 64, 69–70, 81, 157; in art, 133; evidence-based, 238; in PBL, 167; professional development for, 245–46; STEM and, 253; written, 57
Community and School Connections, *234*
community events, 31
community organizations, 267; collaboration with, 183, 223–24
community partnerships, 93–94; for PL, 267; place-based learning relation to, 268
community power, 296
community science-driven plan, HAB and, 11
compost, 30; in cafeterias, 65, 85, 229–30
computer science, 76
Computer Science and Design Thinking, 76
Concord Consortium, 106
Conference of Paris (COP), 256
Conference of Youth (COY), 256–57
Connecticut, 285; Environmental Literacy Plan of, 289; underserved communities in, 290
Connecticut State School Board, 291
Connecting on Climate, 246
Conservation Good Turn certificate program, 240
conspiracy theorists, 159, 238
consumption, 251
contamination, of drinking water, 10
content understanding, data visualization for, 102
controlled burns, 9–10, *10*
"controversial issues," 195–96
controversial questions, 119
Coon, Bryce, 3
COP. *See* Conference of Paris
coping strategies, 181, 183, 186, 188; maladaptive, 185, 189
Cordero, Eugene C., 1–2
"core ideas," 76
corporate philanthropy, 280
Cortada, Xavier, 144
Corwin, Jeff, 233
counselors, mental health and, 182

COVID-19 pandemic, 1, 68, 195–96, 231, 282; field trips relation to, 87
COY. *See* Conference of Youth
credibility, of resources, 212, 220
critical consciousness, 134; for meaning-making, 195
"critical pedagogies of place," 194
critical thinking, 41, 43, 46, 47; in bicycle model metaphor, 54–55; collaboration in, 63; data literacy and, 101; hopefulness relation to, 78; inquiry and, 118; knowledge and, 56, 57–58; large-scale data collection and, 49; in middle schools, 77; PBL and, 166, 168; in research-based activities, 154, 158; resources for, 213; in secondary education, 54
cross-curricular learning, 50; for pre-K, 205
cross-disciplinary connections, 271
Crowley, Kevin, 54
CRT. *See* culturally responsive teaching
CSU. *See* California State University
culturally responsive teaching (CRT), 62; for low-income families, 219; resources for, 212, 220
cultural roots, EcoJustice and, 194
curriculum revisions, 86
cyanobacteria, 10–11

dance performances, 141
DAP. *See* developmentally appropriate practice
data analysis, 111; data visualization for, 100; for decision-making, 99; predictions and, 110
data literacy, 44; critical thinking and, 101; manipulatives for, 103
data moves, 101–2
data ownership, 100–101
dataset sizes, 110
data visualization, 80, 100, *104*, *109*, 111; CODAP for, 106; for content understanding, 102; in elementary schools, 108; executional tasks for, 101; in high schools, 109; PBL and, 167; for precipitation, 106, *107*; sensemaking of, 110; Tableau Data Kids for, 103

DBAE. *See* Discipline-Based Arts Education
debriefing stage, for PBL, 173
decision-making, 45, 67, 95; collective, 54; data analysis for, 99; emissions and, 87–88; resources for, 213; science in, 116; SSI and, 119, 129
Delaware River, 12–13
Delaware Tribe, 196
Denes, Agnes, 135
design-based approach, for PL, 265
design challenge guidelines, *170*
designed environments, 237
design thinking, 169, 171; collaboration in, 90, 167; critical thinking and, 168; NJDOE and, 165; systems thinking and, 176
developing nations, 137
developmentally appropriate practice (DAP), 27; in early childhood education, 30–31; specialist teachers relation to, 84–85
diatoms, 12
digital tools, for collaboration, 214
dimensions, for data analysis, 99–100, 111
disabilities, 62
disciplinary content, 265–66
Discipline-Based Arts Education (DBAE), 138
displaced communities, 27
district administrators, 85
diversity, 95
drinking water, 207; contamination of, 10; service-learning and, 93
drought, 10; precipitation relation to, 13
Dumping in Dixie (Bullard), 134
Dziedzic-Elliott, Ewa, 154

early childhood education, 28; DAP in, 30–31; EcoJustice in, 199, 201–2; environmental literacy in, 29–30. *See also* elementary schools
Earth, 36, 198
Earth Day, 134, 158
Earth Day Education Fund, 280
Earth Friends, 28, 32–36
"Earthrise," 141
ECCLPs. *See* Environmental and Climate Change Literacy Projects

EcoAmerica, 246
eco-anxiety. *See* climate anxiety
eco-charger bicycle, 92
eco-grief, 180
EcoJustice, 193–94, 196, 200; in early childhood education, 199, 201–2; place-based learning and, 195, 207; for pre-K, 205–6; reflection and, 197–98, 201; state standards and, 206
Ecological City, 141
ecological transition zone, 14–15
EcoSchools U.S., 240–41
ecosystems, 12–13; evolution of, 243; populations relation to, 121; resilience of, 202
ECOVIM 250, 230
Education Development Center, 283
education for action (EfA), 66; solutions-oriented perspective and, 68
education for climate action (EfCA), 66–67
education for sustainability (EfS), 194–95
Education Week (magazine), 3
Edutopia, 173
EdWeek Research Center, 28
EfA. *See* education for action
EfCA. *See* education for climate action
EfS. *See* education for sustainability
Egg Harbor High School, 242
elementary educators, 20, 50, 51
elementary schools, 88; Bear Tavern, 224, 233; data literacy in, 103; data visualization in, 108; green spaces in, 288; makerspaces in, 89–90; Primary Search for, 156; PTSD in, 180; service-learning in, 93; Slackwood, 43–44, 51; visual reminders in, 91
Eliasson, Olafur, 137
emissions, 87–88
Emissions Gap Report, of IIPCC, 131
emotional awareness, 181; regulations and, 187, 190–91; resilience and, 186
emotional needs, 182
emotion-focused coping, 185, 188
emotions, 67–68, 145, 186–87, 245; in art, 146; environmental action relation to, 55; teachers relation to, 181–83; verbalization of, 190

empathy: art and, 132, 147; for climate anxiety, 187; of informal educators, 244, 245; outdoor learning for, 45
Endangered Species Act, 43
endangered species art, 142–43, 145
Endeavor STEM Teaching Certificate Project, 270
"enduring understandings," 76
energy bill, 65
energy industry, 167
energy production, 263
engagement, 173; mental health relation to, 189; resources for, 212
engineering content, 76
English-language learners, 211, 219
Ennes, Megan, 263
environmental action: emotions relation to, 55; in families, 252
Environmental and Climate Change Literacy Projects (ECCLPs), 286–88, 289
environmental art, 134
environmental awareness, 31, 37
environmental collage, *132*, 142, 145
Environmental Education Act (1971), 279
Environmental Education Council, 279
Environmental Education in New Jersey (NJCEE), 279
environmental justice, 83, 287; art for, 133, 134, 140, 141; in California, 288–89; SEL and, 87
environmental literacy, 39, 279, 287, 292; in early childhood education, 29–30; No Child Left Behind Act relation to, 280; rain barrels for, 37–38; in secondary education, 53; in social studies, 291
Environmental Literacy Plan, of Connecticut, 289
environmental lobbying, 256
environmentally conscious citizens, 51
environmental monitoring projects, 45
Environmental Protection Agency (EPA), 78, 134, 202. *See also* New Jersey Department of Environmental Protection
environmental racism, 135, 181
environmental values, 252

EPA. *See* Environmental Protection Agency
equity, 174
erosion, 11–12; rainfall relation to, 13; watershed organizations relation to, 232
Essex County, *107*, 110
ethical considerations, 53, 121; for mitigation, 128; perspective in, 127; in SSI, 119
European Commission, 22
evidence-based argumentation, 129; PBL and, 168
evidence-based communication, 238
evolution, of ecosystems, 243
executional tasks, for data visualization, 101
experience-based learning, 30, 65; games for, 239
experts: as resources, 214–15, 220; subject-area, 106; teachers relation to, 58
extreme weather-related events, 179; sea level rise relation to, 218

fact-based foundational guide, 86
fact checking, 159
Fairey, Shepard, 143
false information, 31
families, 86, 88, 89; environmental action in, 252; low-income, 211, 219; in nature, 251; pro-environmental behaviors in, 249–50; SEL in, 94–95
family vacations, 251
farmers, 224; resources from, 234
farmers' markets, 228–29
"Feels Like Summer," 141
Feldman, Allan, 262
fictional books, 89
field trips, 87, 216–17; to beaches, 250; at Haneman Environmental Park, 241
Fifth National Climate Assessment, 144, 288
Finland, 155
First Amendment, 122, 125–26
fixed-mindset, 88, 89; growth-mindset *versus*, 88
Flint Is Family (photo series), 137
Flint water crisis, 137
flood funds, for sea level rise, 68

food, 123, 207, 228–29; from gardens, 227; greenhouse gases relation to, 64
food security, 268
food waste, 93; in cafeterias, 229–30
Forest Fire Service, of New Jersey, 9
Forest Resource Education Center, 226
fossil fuels, 7–8; power plants and, 9
foundational knowledge, 213, 220, 266; for digital tools, 214
"four corners" discussion, 122, 128
four-step inquiry arc, 116
Frasier, Ruby LaToya, 137
Fridays for Future, 137, 281, 292–93
funding, 168; board of education and, 96; for PL, 272; for professional development, 85, 292, 294
FY2000 Integrating the Environment as a Context for Learning grants, 280

Gablik, Suzi, 134
Gallup poll, 196
games, for experience-based learning, 239
Garden for Wildlife, 242
gardens, *225*; compost for, 230; farmers' markets relation to, 228; outdoor learning in, 93, 226, 227; pollinator, 218, 221n10, 225–26, 232; sustainability and, 229
Garoian, Charles, 137
gateway experiences, 245
Geiger, Nathaniel, 238
George Mason University's Center for Climate Change Communication, 238
Girl Scouts of the USA (GSUSA), 239–40
The Giving Tree (Silverstein), 205
glaciers, 137
Global Climate Strike, 281
The Global Learning and Observations to Benefit the Environment Program (GLOBE), 104; Atmosphere Investigation of, *105*
global warming, 28, 131, 245
Global Warming Stripes, *139*
Global Youth Statement, 256
GLOBE. *See* The Global Learning and Observations to Benefit the Environment Program

Golden-Mantled Tree Kangaroo, *133*
Goldsworthy, Andy, 139
Google Earth, 217
Gore, Al, 135
Gorman, Amanda, 141
grants, 85, 95, 301; for compost, 229; FY2000 Integrating the Environment as a Context for Learning, 280; for librarians, 159; from National Science Foundation, 270; from Sustainable Jersey for Schools, 230
graphic organizers, 124, 128
Gravity Hill Farm, 224
green economy, 1
Green File database, 156
greenhouse, for outdoor learning, 224, 227
greenhouse gases, 59; food relation to, 64; heat waves relation to, 9; Industrial Revolution relation to, 7–8; Kyoto protocol and, 251
Greenland glaciers, 137
green skills, 31
green spaces, 200; in elementary schools, 288
Green Team, 85, 96, 254
Grist Magazine, 135
ground-level ozone, 9
group-think dynamics, 169
growing practices, sustainability in, 216–17
growth-mindset, 88, 89; fixed-mindset *versus*, 88
Gruenewald, David A., 194
GSUSA. *See* Girl Scouts of the USA
guidance counselors, 92–93
guided meditation, 197
Guyas, Anniina Suominen, 137

HAB. *See* harmful algal bloom
Hammond, Zaretta, 62
"hands-off" approach, 83
hands-on activities, 30
Haneman Environmental Park, 241
Hansen, James, 134
harmful algal bloom (HAB), 10, 11
Harrison, Helen Mayer, 135
Harrison, Newton, 135
Hayhoe, Katharine, 3

heat islands, 104, 218, 221n10
heat-related injury, 8
heat waves, 8; greenhouse gases relation to, 9
Hersey, Trisha, 184
Hicks, Laurie, 136
hierarchical systems, 194
higher education institutions, 285, 286
higher-order thinking (HOTS), 42, 47
Highlands region, 7
high schools, 301; data visualization in, 109; EfA in, 66; environmental justice in, 83; environmental literacy in, 53
High Water Line (Mosher), 135, 136
Hoffman, Elizabeth, 135
holistic understanding, 56
hopefulness, 43, 67–68, 187–88; critical thinking relation to, 78
horseshoe crabs, 242
HOTS. *See* higher-order thinking
House Bill 5285, 290
"how" questions, 42–43
the Hub. *See* New Jersey Climate Change Education Hub
human behavior, 86
human-environment interaction, 124
Human Impact Project, 61, 63; in middle schools, 65
humanities, 54
humidity levels, 8; air quality relation to, 9
Hurricane Katrina, 135
hydroelectric dams, 219
hydrologic systems, 13

IFLA. *See* International Federation of Library Associations and Institutions
IL. *See* Information Literacy
implementation: of EcoJustice, 200–201; of PBL, 172
inclusivity, 212
An Inconvenient Truth (film), 135
independent learning, 62
Indigenous people, 196, 244; biodiversity relation to, 142
individual actions, 55
Industrial Revolution, greenhouse gases relation to, 7–8
industry engagement, 267

informal education organizations, 223
informal educators, 243, 244–45
informal environments, 237
Information Literacy (IL), 154
innovation, 235
Innovations in Climate Education, of NASA, 262
inquiry: collaboration in, 176; critical thinking and, 118; in social studies, 116–17
inquiry-based learning, 173; for teachers, 269–70
Instagram, 255
integrated approach, 51
"integrating data into teaching," 102, 109, 111
interconnectedness, 54, 67
interdependency, 194
interdisciplinary collaboration, 56, 66–67, 70; NJDOE and, 120; PBL and, 59
interdisciplinary education, 23, 24, 53–54, 281–82; in art, 140; in middle schools, 59; NJDOE and, 84, 95; in PL, 262; professional development for, 59; resources for, 220; in science classrooms, 56–57; in social studies, 57, 81; in STEM, 75–76
Interdisciplinary Standards Alignment Map, 60
intergenerational learning, 235, 249
Intergovernmental Panel on Climate Change (IPCC): *Emissions Gap Report* of, 131; *Special Report on Global Warming of 1.5 Degrees Celsius* of, 133
International Federation of Library Associations and Institutions (IFLA), 153
International Technology and Engineering Education Association, 75
invasive species, 13–14; phenomena-based instruction and, 79; in pollinator gardens, 232
Inwood, Hilary, 134, 137–38
IPCC. *See* Intergovernmental Panel on Climate Change
irrigation, 13
Italy, 4

Jersey Clicks, 156
journals, 47, 183
justice: environmental, 83, 87, 133, 134, 140, 141, 287, 288–89; social, 199, 200, 205

Kean, Thomas H., 279
Keva planks, 160
Khan Academy, 122
kindergartners, 86; list activities for, 47–48; Slackwood Climate Kids for, 44
King, Roger, 136
knowledge, 261, 262, 264; collective efficacy and, 54; critical thinking and, 56, 57–58; in early childhood education, 28; foundational, 213, 214, 220, 266; practical action relation to, 66; of teachers, 23
koala crossing, *91*
Krug, Don, 135
kudzu vine, 14
Kyoto protocol, 251

Lancet global study, 145–46
landfills, 229, 230; storms relation to, 241
land use, 123–24
language arts classes, 54
large-scale data collection, 48; collaboration for, 49–50
large-scale mitigation, 241
Lawrence Township, 85
Lawrenceville Elementary School, 88
Lawrenceville Main Street Organization, 93
LCOYUSA. *See* Local Conference of Youth
leadership skills, 217
"lead learners," 216
learner-centered approaches, 62
"Learning in Places," 268
LEED-certified building, 61, 65
Legislative Document 1902, 295, *295*, 296
Lenape people, 196
lesson plans, 31
librarians, 90, 153, 161; grants for, 159; teachers relation to, 155, 157–58, 160
library materials, 156; weeding of, 157, 162n13
Library of Things (LoT), 155–56, 157

library sustainability, 155
life skills, 213
Lin, Maya, 137, 142
list activities, 47–48
literacy, 117, 140, 218, 288, 289; data, 44, 101, 103; environmental, 29, 37–38, 39, 53, 279, 280, 287, 291, 292; information, 154; in NGSS, 59, 61; scientific, 57, 115–16, 282
A Living Time Capsule-11,000 Trees, 11,000 People, 400 Years (Denes), 135
living wall, 230–32
Local Conference of Youth (LCOYUSA), 256
localized initiatives, 223
logistical planning, for PL, 268–69
LoT. *See* Library of Things
LOTS. *See* lower-order thinking
Louv, Richard, 46
lower-order thinking (LOTS), 47
low-income families, 211; CRT for, 219
Luntz, Frank, 245

MADE CLEAR project, 270–71
Madhubani art, 143
Maine, 285, 292–93, 296; Climate Education Hub of, 295
Maine Climate Council, 293
Maine Environmental Education Association, 293–94, 295
Maine Environmental Priorities Coalition, 294
Maine Youth for Climate Justice, 294
maker-carts, 90, 91
makerspaces, 89–90, 91, *92*, 156–57; in cafeterias, 171; collaboration in, 159–60
Making the Grade? How State Public School Standards Address Climate Change (report), 299, 300–301
maladaptive coping strategies, 185, 189
manipulatives, for data literacy, 103
Mapping Eco-Art Education (Inwood), 137–38
marginalized communities, 286–87
marginalized identities, 180–81
Martusewicz, Rebecca A., 194
Massachusetts, 299

mass media, 262
mathematics, 84; claim-based statements in, 77; Common Core State Standards for, 56–57. *See also* science, technology, engineering, and mathematics
McCaffrey, Mark S., 266
McKibben, Bill, 135
meaning-focused coping, 185–86, 188
meaning-making: art for, 140; critical consciousness for, 195; reflection and, 171
media specialists, 89
meditation, 197. *See also* mindfulness
medium-order thinking (MOTS), 47
Melting Ice Cube Race (activity), 35
mental health, 131, 179; counselors and, 182; eco-grief relation to, 180; engagement relation to, 189; nature relation to, 184
Mercer Meadows, *10*
metacognition, 154, 167; for self-assessment, 171
Miami University of Ohio, 254
Mid-Atlantic Climate Change Education Collaborative, 271
middle schools, 69, 104; bicycle model metaphor in, 61; "core ideas" in, 76; critical thinking in, 77; data visualization in, 80, 108; EfA in, 66; environmental justice in, 83; environmental literacy in, 53; Human Impact Project in, 65; interdisciplinary education in, 59; NGSS in, 79; PBL in, 63; social studies in, 81; Watershed Institute relation to, 239
Middle Search, 156
Mills, Janet, 295, *295*
mindfulness, 189; EcoJustice and, 195
misinformation, 57
mitigation, 125, 127; ethical considerations for, 128
Montclair State University, 269
Moore, Tara, 138
more-than-human world, 131, 139, 145; art and, 132, 133, 134, 146–47, 149
Morris, Nyombi, *135*, 143
Morven Museum, 93–94
Mosher, Eve, 135, 136

motivation, 62, 63; PBL and, 64; for student engagement, 77; teachers relation to, 69–70
MOTS. *See* medium-order thinking
Movement for Black Lives, 133, 137
Mr. John & the Polar Bear (activity), 34–35, 38
Mural Arts Philadelphia Climate Justice Initiative, 141
Murphy, Tammy, 1, 281, 300
MyCoast, 245

NAAEE. *See* North American Association for Environmental Education
NASA, 78, 215; Innovations in Climate Education of, 262; Office of STEM Engagement of, 270
NASA Next Gen STEM For Educators, 103
Nation, Molly Trendell, 262
National Art Education Association, 135, 136
National Center for Science Education, 299
National Council for the Social Studies (NCSS), 117–18
National Council of Teachers of Mathematics, 22
National Earth Month, 94–95
National Geographic, 22
National Institute of Environmental Health Science, 239
National Network for Ocean and Climate Change Interpretation (NNOCCI), 238
National Oceanic and Atmospheric Administration (NOAA), 50–51, 262; Climate Literacy Guide and, 288
national parks, 251, 253
National Public Radio (NPR), 285
National Science Foundation, 29, 270
National Science Teachers Association, 3, 22, 283
National Wildlife Federation, 240–41
native plant species, in living wall, 230–31
native wildlife, 226
natural disasters, 27
natural hazards, 79–80
natural materials, in art, 139

INDEX 313

nature, 231, 251, 257; mental health relation to, 184
Nature-Based Education Consortium, 293
Nature Center of Cape May, of NJA, 242
NCSS. See National Council for the Social Studies
Newark, redlining in, 106, *107*
New Jersey. *See specific topics*
New Jersey Audubon (NJA), 221n10, 240, 242, 279
New Jersey Climate Change Education Hub (the Hub), 19, 22, 23–24, 120, 123
New Jersey Commission on Environmental Education (NJCEE), 279
New Jersey Core Curriculum Content Standards, 279–80
New Jersey Department of Clean Energy, 166–67
New Jersey Department of Education (NJDOE), 20, 280–81; design thinking and, 165; interdisciplinary collaboration and, 120; interdisciplinary education and, 84, 95
New Jersey Department of Environmental Protection (NJDEP), 1, 11, 15, 144, 279; ozone relation to, 9
New Jersey Education Association, 94
New Jersey Forest Fire Service, 9
New Jersey Invasive Species Strike Team, 14
New Jersey: Protect What You Love, 143–44, 145
New Jersey School Boards Association, 2
New Jersey State Bird, *136*
New Jersey Student Learning Standards (NJSLS), 56–57, 133; for science, 117; for social studies, 118, 168
Next Generation Science Standards (NGSS), 56–57, 84, 117, 217, 292, 299–300; in Connecticut, 289; GLOBE and, 104; literacy in, 59, 61; in middle schools, 76, 79; performance expectations in, 2; phenomena-based instruction and, 79; PL and, 266
NGSS. See Next Generation Science Standards
Niche, 201
Nicholls, Jennifer, 155

Niederer, Jess, 224
Niepold, Frank, 283
NJA. See New Jersey Audubon
NJCEE. See New Jersey Commission on Environmental Education
NJDEP. See New Jersey Department of Environmental Protection
NJDOE. See New Jersey Department of Education
NJSLS. See New Jersey Student Learning Standards
NJ Student Climate Challenge, 242
NNOCCI. See National Network for Ocean and Climate Change Interpretation
NOAA. See National Oceanic and Atmospheric Administration
NOAA Global Monitoring Lab Mauna Loa, 102
No Child Left Behind Act (2008), 280
nonformal education, 237, 238, 239–40
nonprofits, 223, 267; NJA, 221n10, 240, 242, 279; SubjectToClimate, 19, 20, 23, 217, 243, 292, 295; Watershed Institute, 232–33, 238–39
non-STEM data, 106, 108
nontraditional educators, 199
North American Association for Environmental Education (NAAEE), 3, 29, 237, 264
NPR. See National Public Radio
Nyombi Morris, Climate Hero, *135*

OASIS. See Organizing Action on Sustainability in Schools
observation checkpoints, for outdoor learning, 45–46, *46*
ocean acidification, 12, 79
ocean temperatures, 12
Office of Climate Action and the Green Economy, 1
Office of STEM Engagement, of NASA, 270
Ohio State University, Art & Ecology Colloquium at, 135
oil infrastructure, 137
online communities, 255
operational barriers, 55, 63, 65, 68, 69

Organizing Action on Sustainability in Schools (OASIS), 242
Outdoor Equity Alliance, 224
outdoor learning, 45–46, *46*, 224, 268; in gardens, 93, 226, 227; large-scale data collection and, 48
ozone, 8–9, 251–52

Pakistan, 147
Palm, Christine, 289–90, 291
parent organizations, 195
parents, 225
parent-teacher organizations, 234
Paris Agreement, 216
Paris Climate Accord, 136
participatory research, 194–95
patience, 88, 246
PBL. *See* problem-based learning
PCK. *See* pedagogical content knowledge
PDSN. *See* Professional Development School Network
pedagogical content knowledge (PCK), 261, 262, 264
Pelto, Jill, 144
People, Places and Environments, in NCSS, 118
People's Climate March, 137
Peppermint Tree Child Development Center, 37–38, *39*
perfectionism, 245
performance expectations, in NGSS, 2
perspective, 62, 67–68, 218–19; bias in, 212–13; in ethical considerations, 127
perspective-taking skills, 128–29
phenomena-based instruction, 79
photo-elicitation, 205
photovoice project, 202–3, *204*, *205*; place-based learning and, 204
Piaget, Jean, 29–30
picture books, 218
Piedmont region, 7
Pine Barrens, 7, 9
PL. *See* professional learning
place-based learning, 77, 78, 267–68; EcoJustice and, 195, 207; at national parks, 251; photovoice project and, 204; for PL, 271

planning stage, for PBL, 172
poetry/spoken word, 141
Points of View, 156
polar zone, 37
political views, 86, 263
politicization, 282
pollinator gardens, 218, 221n10, 225–26; invasive species in, 232
population dynamics, sea level rise and, 242
populations, ecosystems relation to, 121
position statement, 127
positive engagement, 62
post-traumatic stress disorder (PTSD), 180
Pötsönen, Ulla, 155
power plants, fossil fuels and, 9
practical action, 66
"Prayer to Mother Earth," 196–97
precipitation, 8, *11*; data visualization for, 106, *107*; drought relation to, 13; rain barrels for, 37–38
predictions, data analysis and, 110
pre-K, 205, 206
Preparing for a Green and Blue Workforce, of Education Development Center, 283
Primary Search, 156
PRISM Center, of Montclair State University, 269
problem-based learning (PBL), 57–58, 62, 68–69, 70, 165; collaboration in, 63, 167; critical thinking and, 166, 168; debriefing stage for, 173; interdisciplinary collaboration and, 59; professional development and, 172; reflection for, 174; in STEM, 76–77; systems thinking and, 176; UDL and, 64
problem-solving, 41, 43, 47, 57–58, 166; art relation to, 132; collaboration in, 63, 167, 264; outdoor learning for, 45
pro-environmental behaviors, 249–50, 257
professional development, 21–22, 23–24, 155; CMSCE and, 175; for communication, 245–46; ECCLPs and, 286; EcoJustice, 193, 195–200; funding for, 85, 292, 294; for interdisciplinary education, 59; PBL and, 172; for specialist teachers, 84–85

Professional Development School Network (PDSN), 19
professional learning (PL), 261, 262–64, 266, 285; community partnerships for, 267; design-based approach for, 265; funding for, 272; logistical planning for, 268–69; MADE CLEAR project and, 270–71; place-based learning for, 271
professional learning design framework, 264, *265*, 269, 272
project-based learning, 80
Project Dragonfly, 254
Project Drawdown, 1–2
Project Sustainability, 230
Project WET, 239
Providing Hope, 227
PTSD. *See* post-traumatic stress disorder
Public Act 21-2ss, 291
public awareness, 134, 279
public educational art, 242
public libraries, 154–55, 160–61
public policy issues, 122
Purple (film), 137
Puttick, Steven, 262

Quebec Fuentes, Sarah, 263
questioning skills, 47

racism, environmental, 135, 181
radical thinking, 148–49
rain barrels, 37–38, 227
rainfall, 10, *11*, 13, *104*. *See also* precipitation
read-alouds, 205
"real-world" resources, 214
recycling, 30, 44, 88, 241; in cafeterias, 65, 85; service-learning and, 93
redlining, in Newark, 106, *107*
The Re-Enchantment of Art (Gablik), 134
reflection, 47, 173, 195, 203, 216; EcoJustice and, 197–98, 201; meaning-making and, 171; for PBL, 174; in problem-solving, 167; on sustainability, 69
regulations, 167, 218–19, 221n11; emotional awareness and, 187, 190–91; of ozone, 8–9

renewable energy options, 87, 252
Report on K-12 Climate Change Education Needs in New Jersey, 2–3, 29
research-backed storytelling, 296
research-based activities, 89, 124, 128, 267; critical thinking in, 154, 158
resilience, 39, 43, 125, 186, 202. *See also* climate resilience
Resilient Schools and Communities (RiSC), 241–42
resource allocation, 174
resources, 22, 157, 211, 215, 216; for CRT, 212, 220; from farmers, 234; state standards relation to, 213; for student-centered learning spaces, 214, 221
rest, 184
Rest is Resistance (Hersey), 184
reuse: in art, 139; at farmers' markets, 228; LoT and, 155–56
Revival Field (sculpture), 134
Reyes, Geraldo, 290
Ridge and Valley region, 13
RiSC. *See* Resilient Schools and Communities
role models, 182
Rolling Harvest, 224
Rosenello, Patrick, 242
Rosing, Minik, 137
Rotten, A Sweet Deal (film), 195
Rutgers Cooperative Extension, 221n10, 229
Rutgers University, 106, 166, 175, 242

saltwater ecosystems, 12
Sams, Doreen, 138
Sams, Jeniffer, 138
Sandy Hook beach, 250
San Joaquin Valley, 289
Saving Us (Hayhoe), 3
scales, for data analysis, 99–100, 111
school administrators, 87, 95, 266
School Education Gateway, 22
school ground cleanup, 233–34
school walkouts, 281
schoolyard pond, *226*
science: collaboration in, 116; EcoJustice and, 206; library materials on, 156;

NJSLS for, 117; social studies and, 121–26, 127–29, 207; SSI for, 115; state standards for, 299–301
Science, Technology, and Society, in NCSS, 118
science, technology, engineering, and mathematics (STEM), 103, 270; communication and, 253; interdisciplinary education in, 75–76; PBL in, 76–77; SDGs relation to, 80–81
science classrooms, 53; claim-based statements in, 77; interdisciplinary education in, 56–57; PBL in, 63
science communication, 69–70
science experiments, 30
science literacy, 57, 115–16, 282
scientific method, 115
scoring rubric, 126, *126*, 127; for evidence-based argumentation, 129
Scouting America, 240
SDGs. *See* sustainable development goals
sea level rise, 13, 44, 79, 217; extreme weather-related events relation to, 218; flood funds for, 68; population dynamics and, 242
sea walls, 168
secondary education, 53, 54. *See also* high schools; middle schools
sedimentation, 11–12
SEL. *See* social-emotional learning
self-assessment, metacognition for, 171
self-directed inquiry, 58
sensemaking, of data visualization, 110
service-learning, 92–93
Sezen-Barrie, Asli, 67
Sinek, Simon, 235
"Six Americas," 238
Slackwood Climate Kids, 44–45
Slackwood Elementary School, 43–44, 51
SLIDE study, 154
small-group dynamics, 169
Smokey the Bear, 166
social-emotional learning (SEL), 27, 86–87; in early childhood education, 30; in families, 94–95
social justice, 199, 200, 205
social media, 255–56

social studies, 54; civic action and, 120; claim-based statements in, 77; climate activism and, 78; EcoJustice and, 206; environmental literacy in, 291; inquiry in, 116–17; interdisciplinary education in, 57, 81; NJSLS for, 118, 168; science and, 121–26, 127–29, 207; SSI for, 115
social studies literacy, 117
socio-economic status, 220
Socioscientific Issues (SSI), 115; decision-making and, 119, 129; NJSLS and, 118; research-based activities and, 128
solar panels, 65
solar system, 33
solutions-oriented perspective, 67, 68, 218–19
Sonkkanen, Leila, 157
special education, 59
specialist teachers, professional development for, 84–85
Special Report on Global Warming of 1.5 Degrees Celsius, of IPCC, 133
Spencer, Herbert, 233
Spildooren, Oliva, 232
SSI. *See* Socioscientific Issues
stakeholder meetings, 172
STAMP. *See* Standards Transparency and Mastery Platform
standards-based content, 263
standards-based curricular culture, 207
standards-based resources, 211, 216
Standards Transparency and Mastery Platform (STAMP), 84
Standing Rock, 133
state climatologist, 214
State Hazard Mitigation Plan, 12
state standards, 201; EcoJustice and, 206; resources relation to, 213; for science, 299–301. *See also* New Jersey Student Learning Standards
statistical thinking moves, 110, 111
STEM. *See* science, technology, engineering, and mathematics
Stevenson, Robert B., 155
stewardship, 253

storms, 44; aquatic ecosystems relation to, 11–12; landfills relation to, 241; watershed organizations relation to, 232
Stream Watch Schools, 232–33
student advocacy, 215
student-centered learning spaces, resources for, 214, 221
student-collected data, 100–101; GLOBE and, 104
student engagement, academic achievement relation to, 77
student interests, 267
student leaders, 235
student mindset, 173
student readiness, 174
Studies in Art Education (journal), 136, 137
subject-area experts, 106
SubjectToClimate, 19, 20, 23, 217, 243, 292; Climate Education Hub and, 295
summer camps, 229
summer enrichment, 228
sunlight observation centers, 42
Sunrise Movement, 290
Superstorm Sandy, 12, 135, 136, 241; Bradley Beach and, 168
Surface Temperature data collection protocol, 104
surface water, HAB and, 10, 11
sustainability, 61, 153, 194–95, 202, 207, 240–41; audits for, 175; Climate Change Council for, 158–59; gardens and, 229; Green Team and, 96; in growing practices, 216–17; library, 155; reflection on, 69
Sustainability in Libraries (ALA), 154
Sustainability Institute, at The College of New Jersey, 2
sustainable development, 94
Sustainable Development Goals (SDGs), 59, 61, 63–64; climate activism and, 78; STEM relation to, 80–81
sustainable habits, 31, 43, 51
Sustainable Jersey for Schools, 94, 229, 230, 242
systems thinking, 47, 76; critical thinking and, 57; non-STEM data and, 108; PBL and, 176
system wide solutions, 106

Tableau Data Kids, 103
Talks, Isobel, 262
TCNJ. *See* The College of New Jersey
teacher attrition, 282
teachers, 53, 63, 67, 201, 216, 299; collaboration with, 89, 171, 301; emotions relation to, 181–83; experts relation to, 58; grants for, 95; inquiry-based learning for, 269–70; knowledge of, 23; librarians relation to, 155, 157–58, 160; motivation relation to, 69–70; PBL and, 166; Slackwood Climate Kids and, 44–45; specialist, 84–85; transdisciplinary approaches of, 148–49
teacher training, 24, 104. *See also* professional development; professional learning
"teaching data skills," 101–2, 109, 111
Teach NJ Act (2013), 280
team-based model, 59
temperate zone, 37
Texas Freedom Network Education Fund, 299
Thanksgiving food drive, 93
"Thinking Set of Tools," 47–48, 49–50
Thunberg, Greta, 64, 92, 137, 143
tidal ecosystems, 13
Tilth Alliance, 268
Tinkergarten, 28
transdisciplinary approaches, of teachers, 148–49
transition day, 169, 171
"trash stash challenge," 69
trauma responses, 179
"Trends in Atmospheric Carbon Dioxide," 102
tropical zone, 37
Tuorto, Steve, 232
2020 Scientific Report on Climate Change, 1

UC. *See* University of California
UDL. *See* universal design for learning
Udo, Nils, 139
ultraviolet radiation (UV), 8
underserved communities, in Connecticut, 290

318   INDEX

UNFCCC. *See* United Nations Framework Convention on Climate Change
United Nations, SDGs of, 59, 61, 63–64, 78, 80–81
United Nations Climate Change pages, 128
United Nations Climate Summit, 119–20, 122–23, 126, 127
United Nations Earth Summit, 134–35
United Nations Framework Convention on Climate Change (UNFCCC), 134–35, 256
United States Global Change Research Program, 288
universal design for learning (UDL), 62, 64
University of California (UC), 286, 288
upcycling materials: in art, 139; weeding and, 157
Urban Heat Island Effect, 104
US Army Corps of Engineers, 241
US Global Change Research Program, 266
UV. *See* ultraviolet radiation

values-based leadership, 240
Vector-Borne, Waterborne and Heat-Related Illnesses activity, 239
venn diagrams, 47
verbalization, of emotions, 190
visual arts, 131, 132–33
visual reminders, in elementary schools, 91
volunteerism, 93
volunteer organizations, 252

WACA. *See* Wildlife and Conservation Association
Walter Cronkite Award for Climate Education, 291
Washington State, 294
wastewater overflow, 241
Water, Climate, and Resilience Guide (Project WET), 239
water crisis, 202
water-loving plants, 232
water regulations, 218–19
water runoff, 202, 203, 204

Watershed Institute, 232–33, 238–39
watershed organizations, 232–33
water wheels, 219
Wazer, Sena, 290
weather, climate compared to, 27–28, 44
weeding, 157, 162n13
West Vancouver Memorial Library, Climate Writer in Residence at, 154
West Virginia Climate Change Professional Development Project, 263
wetlands, 13
What's Missing (Lin), 137
"when" questions, 42
Whitehouse, Hilary, 155
White House Climate Report, 144
White youth, 181
Whitman, Todd, 280
whole-class discussions, 217
"why" questions, 41–42
wildfires, 9, 180
Wildlife and Conservation Association (WACA), 234
Wildwood High School, 241, 242
Windschitl, Mark, 266
wind-turbine blades, 169, *170*, 171
"wonder and notice," 48
word choice, 245–46
Words That Work (Luntz), 245
World Wildlife Foundation, 3
worry, 185
written communication, 57

Xerces Society, 226

Yale Program for Climate Change Communication, 238
Youth and Children's Constituency to the United Nations Framework Convention on Climate Change (YOUNGO), 256
Youth Environmental Society, 254

zine, 42
Zunigha, Curtis, 196

Contributor Biographical Sketches

Marissa Bellino, Ph.D., is an Associate Professor of Education at The College of New Jersey where she teaches social foundations, urban education, and science methods to preservice teachers. Her teaching and research interests explore critical place-based pedagogies, environmental sustainability education, and participatory action research.

Glenn Branch is the deputy director of the National Center for Science Education, a nonprofit organization that defends the integrity of American science education against ideological interference. He is the author of numerous articles on evolution education and climate education, and obstacles to them, in such publications as *Scientific American, American Educator, The American Biology Teacher*, and the *Annual Review of Genomics and Human Genetics*, and the co-editor, with Eugenie C. Scott, of *Not in Our Classrooms: Why Intelligent Design is Wrong for Our Schools* (2006). He received the Evolution Education Award for 2020 from the National Association of Biology Teachers.

Greer Burroughs is an Associate Professor at The College of New Jersey in the Department of Elementary and Early Childhood Education. She received her doctorate degree in Social Studies Education in 2010 from Rutgers University in New Brunswick. Prior to coming to TCNJ, she spent twenty years working in her field teaching students in grades 3 to 12, providing professional development to in-service teachers and teaching at other NJ Universities. Her research interests include educating for democratic and global citizenship and education for sustainability, with a focus on critical place-based learning experiences and participatory research methods. She has brought undergraduate student researchers to Greece and Ecuador to study methods of economic, cultural, and environmental sustainability. This work has led to several publications and conference presentations.

Jim Clifford is the director of Exploring Climate and the State Lead for Subject to Climate's new Connecticut Climate Change Education Hub. He served as an independent contractor with the Connecticut State Department of Education to revise the Connecticut State Social Studies content standards and Social Studies Model Curriculum to include climate change and environmental literacy. Jim also serves on the board of directors for the Connecticut Outdoor and Environmental Education Association. He holds a JD from the Columbus School of Law, Catholic University of America and a BA from Georgetown University.

Eddie Cohen, Ed.D., is the Assistant Director of the Center for Mathematics, Science, and Computer Education at Rutgers University. He holds an Ed.D. in Curriculum and Instruction, MS in School and District Leadership, BA in Elementary Education. His work is around how to design and create K–12 integrated approaches to Climate Change Curriculum using education technology.

Helen Corveleyn has experience in PreK–12 STEM education. She is a STEM teacher at Timberlane Middle School and spent ten years as a STEM facilitator at Hopewell Elementary School. She has a M.A.T. in conservation biology from Miami University of Ohio, which complements her B.S. in environmental policy from Marist College. Helen is the recipient of the Presidential Award in Math and Science Teaching given by the National Science Foundation and the White House. Other honors include receiving the Governor's Educator of the Year Award (2019), the New Jersey County Teacher of the Year (2020), and the "I Can STEM NJ Role Model" for the New Jersey STEM Pathways Network (2021). Most recently, she received the Mercer County Society of Professional Engineers STEM Teacher of the Year Award (2024). Her passion is inspiring young people to become planetary stewards who can communicate scientific ideas and promote innovation in science and sustainability. She teaches graduate students in environmental leadership at The College of New Jersey and in conservation biology at Miami University of Ohio. Her international fieldwork includes studying island biogeography and swimming with whale sharks in Baja, Mexico; studying orangutans and sustainable palm oil in Borneo, Malaysia; and creating a multimedia-based conservation campaign to support the Belize Zoo and Maya Forest Corridor.

Grace Corveleyn is a high school junior from Pennington, New Jersey. Her environmental passions include conservation, marine science, and climate policymaking. She is currently an Ocean Guardian Ambassador for the National Oceanic and Atmospheric Administration (NOAA), and virtually volunteers with The Climate Initiative. She serves as the co-founder and Vice President of Communications of the Youth Oceanic Initiative, a non-profit organization dedicated to making the environmental movement more accessible by spreading the messaging of ocean conservation to students from around the world. She attended the Local Conference of Youth for the United States (LCOYUSA) in 2023, drafting the national youth statement of climate demands that was presented at the 28th Conference of Parties (COP28) last December. In the future, Grace hopes to continue her passion for marine and environmental science through her college education and aspires to pursue a career in marine biology. In addition to her work in the climate space, Grace is a three-sport athlete, avid Girl Scout, and loves spending time outside with her friends and family.

Rachel DiVanno is a middle school science teacher in Metuchen, NJ. She earned her MEd in STEM Education from American College of Education, and BS in Elementary Education and integrated Science, Technology, Engineering and Mathematics at The College of New Jersey. Rachel was one of the first students to earn a minor in Environmental Sustainability Education at TCNJ. She has led the Edgar Middle School Environmental Club and has been a member of the Edgar Middle School Green Team. Rachel's focus is making STEM accessible for all students using an environmental approach.

Dr. Andrea Drewes is an Assistant Professor of Graduate Education, Leadership and Counseling at Rider University where she teaches courses in teacher education. She holds a PhD in Learning Sciences, a MEd in Middle/Secondary Science Education, and a BA in Chemistry Education. She is trained as a learning scientist in mixed methods research methodology and her research focuses on teacher preparation for climate change instruction, student learning outcomes in climate science education, and instructional interventions to facilitate students' scientific explanations. She is the co-editor of the recent book, *Teaching Climate Change in the United States*, which highlights best practices in climate change education through the analysis of a rich collection of case studies that showcase climate change educational efforts across the country.

Ewa Dziedzic-Elliott serves as the subject librarian for all departments in the School of Education. She has ten years of experience as a K–12 librarian, including work in both elementary and high school settings. She holds an MLIS from Rutgers University and an MA in Polish Language and Literature with a minor in Speech Therapy from Jan Kochanowski University, Poland, EU. She also holds NJ supervisor and principal certifications. She is an executive board member for New Jersey Association of School Librarians (NJASL) and currently serves as NJASL president.

Carrie Ferraro, Ph.D., is an Assistant Professor of Practice with the Math & Science Learning Center at Rutgers University. She has fifteen years of experience bringing cutting edge science into the classroom through curriculum development, youth programs, and teacher professional development programming. She is particularly interested in connecting climate change learning to data and local action and has done work on understanding systems thinking around climate change. She is currently the Director of the Rutgers Science Explorer.

Cari Gallagher is a Teacher at Lawrenceville Elementary School in Lawrence Township, the LTPS District Green Team Leader, and a Supervisor of Student Teachers at TCNJ. She concentrates on selecting climate change books that are developmentally appropriate for elementary school students and encourages these resources to be distributed throughout the district. She also works with K–12 teachers to promote sustainable initiatives throughout the district. Throughout observations and conferences with student teachers, she encourages her students to integrate Climate Change resources in all content areas while planning lessons. She holds an M.A in Education and is currently working on a Ph.D. in Curriculum and Teaching.

Kathleen L. Grant, Ph.D., is an Assistant Professor in Counselor Education at The College of New Jersey. She holds a Ph.D. in Counselor Education, an M.A. in Counseling, and a B.S. in Biology. Her research illuminates the experiences of young people as they navigate the climate crisis, specifically focusing on mental health impacts and strategies to support thriving when dealing with eco-anxiety and grief. She has engaged audiences across her region, the United States, and internationally about the nature of the climate crisis and actions to take to support change and positive mental health. Dr. Grant is a Nationally Certified Counselor (NCC), a certified School Counselor, a certified Director of School Counseling Services and is also a grassroots organizer and climate activist.

Ian Gray is the Senior Land Steward for the Mercer County Park Commission in Mercer County, NJ. He earned his B.A. in Ecology and Naturalist Education at The College of New Jersey. Though the focus of his work is on land management on park properties he also educates the public in ways to support native biodiversity and combat climate change with the removal of invasives and planting of native species.

Pat Heaney is the Assistant Director of Education at The Watershed Institute. She is an award-winning educator with over 35 years of experience in non-formal education outdoors. In 2019, she completed a certificate in Climate and Health from Yale School of Public Health, and is now focused on Climate Change Education. She is also helping to revitalize one of the nearly extinct indigenous languages of NJ, the Lenape Unami Dialect.

Eileen Heddy, Ed.D., serves as the Director of the Office of Support for Teacher Education Programs and Global Student Teaching at The College of New Jersey. Her research interests include international field experiences, civic engagement and culturally sustaining pedagogy. She holds a master's degree in history from Rutgers University and an EdD from the University of South Carolina.

Janna Hockenjos is the Founder & CEO of Earth Friends. She holds an MA in Journalism from New York University and a BA in Spanish from Wittenberg University. Her work focuses on closing the climate education gap in early education by providing young children with a foundation of environmental literacy through a content-based curriculum approach to learning.

Missy Holzer, Ph.D., is a Science Curriculum designer at Great Minds PBC, and science education consultant. She holds a PhD in Science Education, MS in Physical Geography, MAT in Science Education, and BS in Environmental Planning and Design. Her work supports curriculum designers, educators, and scientists as they translate science into engaging classroom instructional resources, especially related to climate science and climate change. She is also a Climate Literacy and Energy Awareness Network (CLEAN) Ambassador providing climate change professional development, and she has served on numerous science standards panels for the New Jersey Department of Education.

Kristin Hunter-Thomson, Founder and Director of Dataspire Education & Evaluation LLC, has worked in STEM education for over twenty years in various locations around the United States and abroad. She pulls from her middle/high school science teaching background, her experience in formal and informal STEM education, and her graduate training in science to oversee the growth and development of Dataspire Education & Evaluation LLC. She has extensive experience in facilitating educational professional development for teachers in grades 3–16, in developing or consulting in the development of data-focused educational resources and assessment instruments, and in evaluating educational programs. She holds an M.A.T in Science Education, a M.S. in Marine Science, a G.C. in Program Evaluation, and a B.A. in Biology & Environmental Studies.

Dr. Sami Kahn is Executive Director of the Council on Science and Technology at Princeton University. She holds a Ph.D. in Science Education, a J.D. in Environmental Law, an M.S. in Ecology and Evolutionary Biology, and a B.A.

in Music (concentration Biology). She uses her background in science education and law to inform her research and scholarship on inclusive science practices, socio-scientific issues (SSI), argumentation, and social justice.

Brielle Kociolek, Ed.D., is the Senior iSTEM Coordinator of the Center for Mathematics, Science, and Computer Education at Rutgers University. She holds an Ed.D. in Technology Integration K–16, M.Ed. in Education, a B.S. in Elementary/Special Education and Environment Studies. She is interested in ways to support teachers with implementing iSTEM in their classrooms. Her dissertation focused on researching how to support teachers with technology integration. Dr. Kociolek has fifteen years of experience in K–12 education including designing, implementing, and leading science and technology professional developments and curriculum.

Kelley Le, Ed.D., has been in the educational field for over fifteen years as a high school science educator, instructional coach, and educational leader. She is currently the executive director of the UC-CSU Environmental and Climate Change Literacy Projects (ECCLPs), former director of the UC Irvine Science Project, Friends of the Planet Award recipient (2022), and author of Teaching Climate Change for Grades 6–12: Empowering Science Teachers to Take on the Climate Crisis Thoughts NGSS (2021, 2024). She also serves as a CLEAN advisory board member, and a Climate Reality Corps mentor.

Samantha Lindsay is a current special education teacher in North Jersey. She graduated from TCNJ with her B.S. in Early Childhood and Spanish in 2022, and then again in 2023 with her M.A.T. in Special Education. Samantha worked as Dr. Madden's graduate assistant during her graduate year, helping collect data and mainstreaming communications with stakeholders through her NOYCE grant.

Dr. Timothy Lintner is Assistant Vice Chancellor for Academic Affairs and Carolina Trustee Professor of Education at the University of South Carolina Aiken. He holds a Ph.D. in Social Sciences Education, an M.Ed. in Educational Leadership, an M.Ed. in Special Education, and a B.A. in History. His research focuses on the intersection between social studies, special education, and civic action.

Dr. Lauren Madden is a Professor of Elementary Science Education at The College of New Jersey. She holds a B.A. in Earth Sciences–Oceanography, M.S. in Marine Science and PhD in Science Education. Dr. Madden's work advocates for scientific literacy and the health of our planet through teaching and learning. Her research has been supported by grants from the New Jersey SeaGrant

Consortium, National Science Foundation, and US Environmental Protection Agency. She has written a textbook on Elementary Science Teaching Methods along with more than fifty peer-reviewed journal articles and book chapters. She was named the 2021 Outstanding Science Teacher Educator of the Year by the Association for Science Teacher Education and served as the inaugural iCAN STEM Role Model Award by the New Jersey STEM Pathways Network. In recent years, her work has focused directly on K–5 climate change education, and she was the lead author on the New Jersey School Boards' Association & Sustainable Jersey For Schools' *Report on K-12 Climate Change Education Needs in New Jersey*. Her expertise in climate change education in New Jersey has been featured prominently in many media outlets including the *New York Times*, the *Washington Post*, the *Guardian*, NPR, and the *Star Ledger*.

Carolyn McGrath is a Visual Arts Educator at Hopewell Valley Central High School. She holds an MFA from the Maine College of Art, Post-BA Certificate from Parsons School of Design and The New School, and B.A. in Visual Art, Psychology, and English from Rutgers University. Her work with students explores the transformative and liberatory potential of art and art education.

Dr. Jessica Monaghan is the Assistant Director of STEM in the Program for Teacher Preparation at Princeton University. She holds an Ed.D., in Design of Learning Environments from Rutgers University Graduate School of Education. Dr. Monaghan has over a decade of experience as a middle school science teacher and K–12 Science Supervisor. She is passionate about integrating global and local socio-scientific issues in the science classroom and engaging learners in experiential outdoor learning.

Allison Mulch is the Director of Education at New Jersey Audubon, the state's largest and oldest conservation organization connecting all people with nature and stewarding the nature of today for all people of tomorrow. She oversees the internationally recognized Eco-Schools USA program, the U.S. Department of Education Green Ribbon Schools program, and the Resilient School Consortium (RiSC) in New Jersey. She works with school supervisors and teachers to integrate environmental, sustainability, and climate change education across disciplines to create a more unified approach to teaching and learning. Allison serves on the board of the Alliance for New Jersey Environmental Education, is an appointee for the New Jersey Commission for Environmental Education and is a member of the Climate Change Education Thought Leader Committee and its Climate Change Education Initiative. Allison was previously the Director of the NJ Sustainable Schools Consortium at the Educational Information and Resource Center and served as an 2021–22 Administrator for the Organizing Action on Sustainability in Schools.

Jeanne Muzi is the Principal of Slackwood Elementary School in Lawrence Township, New Jersey. With over twenty years of experience in education, Jeanne's passion is working with children to be creative and curious problem solvers. Jeanne has served as a NOAA Teacher At Sea, National Geographic Grosvenor Teaching Fellow and as a member of the District Green Team.

April Oliver has been an elementary school librarian in Lawrenceville, NJ, for eighteen years. She earned her M.A. in English Literature from the College of New Jersey and M.L.I.S. from Rutgers University. Over the past five years, April has assumed the role of Green Coordinator at her elementary school, meeting with students to create science-based projects, some of which are presented at local sustainable festivals. She has also dedicated the library's Makerspace to building cross-curricular connections, particularly in the area of science, by having students conduct experiments with water and earth materials. She continues to foster young children' love of nature and literature by stocking the library's "Climate Corner" nook with books, magazines, and even plants.

Tina Overman is a STEM Teacher at Bear Tavern Elementary School in the Hopewell Valley Regional School District. She is a Climate Reality Project Leader and Co-Chair of the NJ Chapter's Education Committee. Tina led her school's journey to receive the award for the National Wildlife Federation's Eco-Schools USA Green Flag Award with New Jersey Audubon. Her climate-focused teaching centers around fostering the idea of being an active problem finder, solving the challenges of today while keeping an eye on the impacts of tomorrow.

Janice Parker, Ed.D., is a clinical specialist at The College of New Jersey. She holds a doctorate in Teacher Leadership with a focus area in Early Childhood from Rutgers University. Her work seeks to support student teachers in their clinical practice as they learn how to best support children's learning.

Isabelle Pardew is a Research Associate with SubjectToClimate, where she engages in content creation and studies methods for expanding the reach and impact of climate education. She is currently working on her MS in Communication Studies and has a B.S. in Applied & Industrial Mathematics and an MS in Applied Mathematics, both from Towson University. In addition to her current work, she is an advocate for ornithological societies and bird conservation, establishing Lights Out research groups at the universities she attended. Isabelle's work also centers on communicating environmental issues through illustration.

Beverly R Plein, Ph.D., was a classroom teacher and technology facilitator when she received the national Milken Educator Award. She taught graduate and undergraduate social studies methods courses at Montclair State University. She has an MScEd in Educational Technology and her PhD is in Teacher Education and Teacher Development. Beverly worked at the New Jersey Department of Education as the Director of the Office of Standards at the time when New Jersey was first in the nation to adopt climate change standards across content areas.

Kerry Rushnak is a special education teacher in central New Jersey. Kerry holds a BS in Elementary Education and Integrative STEM from The College of New Jersey and a MAT in Special Education. She worked as a graduate assistant under Dr. Lauren Madden in the Environmental Sustainability Education program. Kerry has helped with the organization and publication of numerous books on educational teaching practices and science education.

Julia T. Sims is the Director of Research and Development at SubjectToClimate. In this position, she focuses on writing case studies on the state of climate-change education and guiding development efforts. She holds a B.A. in Economics from Harvard Extension School.

Sarah I. Springer, Ph.D., is an Associate Clinical Professor for the Seattle University Online School Counseling program. Dr. Springer's passion includes mental health and advocacy support in schools. She spent many years as a school counselor and currently supervises school counselors working towards their LPC credentials. In her private practice, she offers counselor consultation groups and workshops to therapists and K–12 educators, complementing her research and scholarship interests around pre-service counselor pedagogy, shame resilience, and educator wellness in schools.

Sarah Sterling-Laldee is Senior Advisor on Climate Change Education for the New Jersey Department of Education. She holds an M.A. from New York University Steinhardt School of Education in Environmental Conservation Education. Sarah has over twenty years of experience in urban science and STEAM education, and is a passionate advocate for expanding access to science and sustainability education for all children.

Kelly Stone is an Elementary STEM Educator for the Long Branch Public School District. She holds an MS from University of Nebraska in Instructional Technology, Post MA Certification in Environmental Sustainability Education from TCNJ, Supervisions Certificate from Caldwell, and a B.A. in Early Childhood Education from Catholic University. Kelly is passionate about

Climate Change and Environmental Education and works to promote the infusion of both in all content areas throughout her district.

Graceanne Taylor has served as the Education and Outreach Coordinator at Save Barnegat Bay for the past five years. Prior to this role, she received a bachelor's degree in marine science. Graceanne has found an extreme passion for education and communication through her work in the Barnegat Bay watershed as an informal environmental educator building over 10 years of experience, educating and engaging audiences of all ages and sizes.

Chris Turnbull has been a principal for 16 years at the Elementary and Middle School levels. He has bachelor's degrees in Elementary Education and History ('00) and a master's Degree in Educational Leadership from The College of New Jersey ('06). In 2022, Chris was named the New Jersey Visionary Principal of the Year and the National Association of Elementary School Principals' "Nationally Distinguished Principal" from New Jersey. Chris is currently the Principal of Timberlane Middle School in the Hopewell Valley Regional School District in New Jersey. Chris has received the Eco-Schools USA Green Flag Award at Ben Franklin Elementary School and at Bear Tavern Elementary School.

Kimi Waite, Ph.D., is an Asian American educator-activist-scholar with over a decade of teaching experience. She is a 2019 Environmental Education 30 Under 30, recognized by The North American Association for Environmental Education; the 2021 California Council for the Social Studies Outstanding Elementary Social Studies Teacher of the Year; and a 2021 public voices fellow on the climate crisis with The OpEd Project and the Yale Program on Climate Change Communication. She is an Assistant Professor of Child and Family Studies at California State University, Los Angeles.

Margaret Wang is the co-founder and COO of SubjectToClimate. Previously, she was a high school economics and social studies teacher. She has worked in edtech as a product manager as well as at Harvard, conducting research in education policy and sustainability education. She holds an M.Ed. in International Education Policy from Harvard Graduate School of Education and a B.A. with a teaching certification from Princeton University.

Karen Woodruff, Ph.D., is an Assistant Professor at Kean University in the College of Education where she teaches science and math methods courses and technology integration for pre-service educators. Her teaching interests include preparing preservice teachers to attend to interdisciplinary climate-change education. She is engaged in research on supporting educators of all experience levels with the meaningful integration of authentic data, specifically evidence of the changing climate. Dr. Woodruff leads the Endeavor STEM Teaching Certificate Project's mission to support in-service teachers with meaningful data integration and culturally affirming sensemaking opportunities.

Dr. Melissa Zrada is an Associate Professor of Integrative STEM Education at The College of New Jersey. She holds a Ph.D. in Cognitive Science in Education, M.S. in Neuroscience and Education, and bachelor's degrees in Technology Education and Interactive Multimedia. Her research focuses on data literacy and student question-asking.

Dr. Christine Whitcraft is a professor in the Department of Biological Sciences and Director of Environmental Science and Policy (ES&P) at California State University Long Beach (CSULB). Her major area of interest is coastal wetland ecology, focused on the impact of anthropogenic activities on functioning of brackish and salt marshes. Specifically, she investigates restoration strategies, impacts of invasive plants, and climate change–related impacts. She received her B.A. in Biology from Williams College, Williamstown, MA, and Ph.D. in Biological Oceanography from University of California, San Diego, Scripps Institution of Oceanography. She is passionate about and engaged in activities that promote equitable and sustainable practices in climate change education.

Amara Ifeji is a National Geographic Young Explorer and internationally awarded non-profit leader in climate and environmental justice. As the Director of Policy with the Maine Environmental Education Association, she leverages grassroots advocacy and participatory justice to advance research-backed local, state, and federal policy solutions. She served as the lead coordinator for Maine's first Climate Education Summit, mobilized a youth-led movement that spearheaded Maine's more than $2 million climate education program, and serves on the Maine Climate Council as the governor-appointed Youth Representative. She holds a B.A. with honors in Political Science from Northeastern University. As a current graduate student, she is pursuing an MSc in Nature, Society, and Environmental Governance at the University of Oxford, fully funded as a Marshall Scholar.

www.ingramcontent.com/pod-product-compliance
Lightning Source LLC
Chambersburg PA
CBHW071400300426
44114CB00016B/2125